Alys Fowler

Eat What You Grow

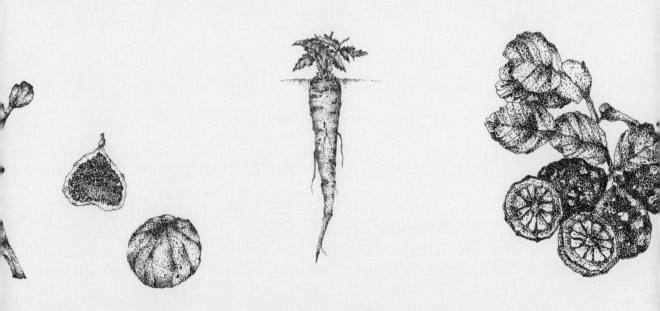

How to have an undemanding edible garden that is both beautiful and productive

AN HACHETTE UK COMPANY

hachette.co.uk

First published in Great Britain in 2021 by Kyle Books
an imprint of Octopus Publishing Group Limited
Carmelite House
50 Victoria Embankment
London EC4Y 0DZ
kylebooks.co.uk

ISBN   978 0 85783 898 8

Distributed in the US by Hachette Book Group, 1290 Avenue of the
Americas 4th and 5th Floors, New York, NY 10104

Distributed in Canada by Canadian Manda Group, 664 Annette St.,
Toronto, Ontario, Canada M6S 2C8

Editorial Director	Judith Hannam
Publisher	Joanna Copestick
Assistant Editor	Florence Filose
Photography	Roo Lewis
Photographic Lab Services	Labyrinth, London
Illustrations	Anka Dabrowska
Graphic Design	Carol Montpart & Alisa Welby
Production	Caroline Alberti

A Cataloguing in Publication record for this title is available from the British Library

Printed and bound in China

10 9 8 7 6 5 4 3 2

Contents

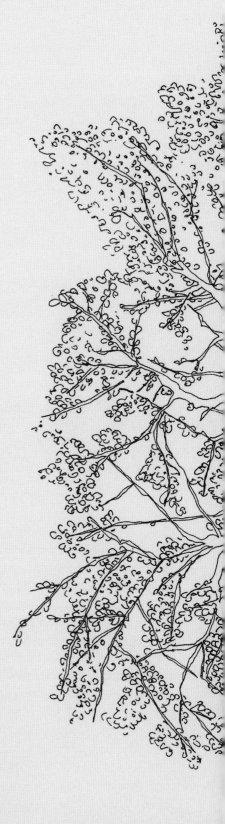

4   INTRODUCTION

What Is Polyculture?
Why Bother?
How This Book Works
In a Nutshell
Some Tips for Sowing and Planting Out Polycultures

25   THE BASICS

Trees
Leafy Greens
Shrubs
Upright Herbs
Climbers
Umbels
Ground Cover
Ornamental Edibles and Pollinator Plants

117   THE FILLERS

133   THE TOPPINGS

182   ADDITIONAL SPACES

The Edible Water Garden
The Window Box
The Shaded Garden

188   INDEX

191   ACKNOWLEDGEMENTS

Introduction

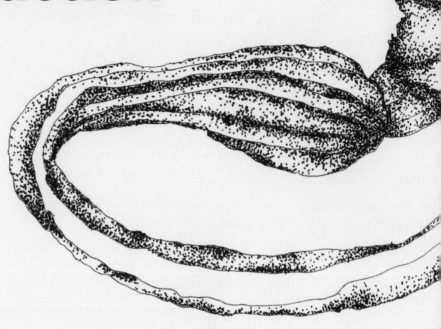

By late summer the garden is set, it is in full motion, the apple trees have branches bowing down in reverence to the bees' work, the final crop of gooseberries is fattening up to a sweetness unparalleled to their former tart selves. The currants, empty of their bounty, are soaking up the sun as summer squash clamber over them like buoyant toddlers to head up into the apple trees for full effect. Fat squash plants sprawl and I have to keep reminding them to be good neighbours to the perilla, kales, chard and lettuce that love basking in their shade.

My many umbels, the Korean celery, angelica, the fennels, dang qui, among the *Ammi majus* and coriander, float around the garden slightly tipsy from the rain, and if I don't get round to harvesting seed they will populate the garden with thousands of babies. The lettuces have all gone to seed, but I like this stage so much, when

they suddenly tower above their companions like strange sculptures, that I've yet to pull them up. This year I experimented with pink celery from China, the stems grow in various shades of raspberry, but they do not like their neighbours or for that matter our varying weather patterns and some, though not all, are flowering in protest. The hoverflies are ecstatic with these flowers and I'm curious about what it takes to please this celery so I'll try again next year.

Lunch is calling, I'll pick a few younger leaves of wild rocket, handfuls of herbs, of parsley, tarragon, marjoram, basil, chives, both garlic and ordinary, with a few mouse garlic flowers, *Allium angulosum,* and I will fold this into an omelette and eat outside gazing upon the glory of a garden so ripe and abundant. For this is my wild bit of world, my tiny slice of reciprocity with nature. The place where I gently tend and

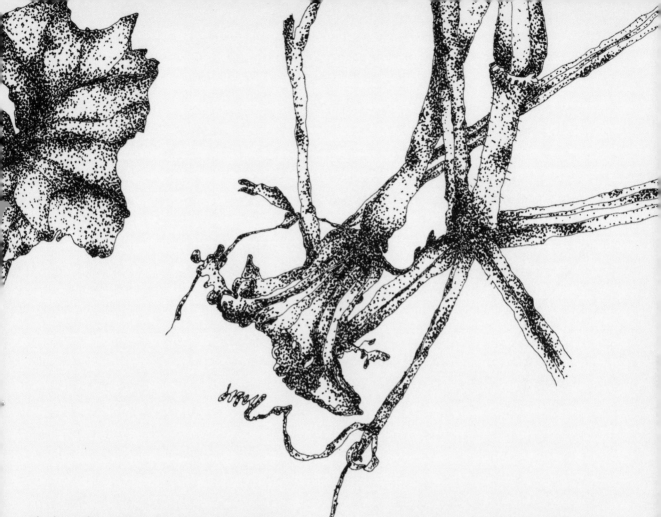

pluck, and in return I and the many others – the beetles, slugs, wasps, bees, birds, microbes, bacteria, mushrooms and so many plants – thrive so that I can call this home.

I'm not one for labelling my garden, in part because it sits quite happily between many categories, but if I had to give it a description it would be this. It's a polyculture of perennial and annual edible plants and herbs that sits on the edge of forest garden ideas, but also borrows from intensive organic principles. On the rare occasion I don't finish my beer, I'll use the leftovers in a slug trap. I have never used any kinds of non-organic chemicals on the garden. I buy in seeds and organic seaweed pellets. I also use liquid seaweed feed and get through around ten 80-litre (17½-gallon) bags of organic, peat-free

compost a year. Other than this and the odd ball of string the garden is a closed system, except for mains water.

The success of the garden is largely down to my compost and I compost very seriously. I'd go as far as saying it's my religion, I love the process of breaking down material. I am fascinated by the many rots, moulds, mushrooms and bacteria that recycle life. It is poetic that the end of everything is the beginning of everything else. Thus, there is no food waste in my house because it is just on the road to becoming food again. The same goes for cardboard, woolly socks, dog hair, manuscripts and anything else that I can get my hands on to feed my compost.

I use three different composting systems at home. I have two Hotboxes for composting garden and kitchen waste. These are black polystyrene bins that raise the temperature of fresh compost quickly to between 50–75°C (122–167°F) meaning that everything cooks down speedily to beautiful black compost. I also have a large rather unkempt compost heap cum wildlife habitat that is mostly made up of thicker material that takes some time to break down. Every year I promise myself I will tidy it up and come up with a more elegant solution and then I don't. Every year a huge bumblebee queen nests in it and I am reminded why tidy is not everything.

I own a green cone, which is a strange green plastic cone with a basket on the bottom that sinks into the ground, and all sorts of rottable stuff can be chucked in it to rot into the ground. I mostly use it for food waste, dog poo and bindweed stems. You don't empty the compost, it just seeps into the soil so it needs to be situated somewhere useful. Mine is next to an 'Egremont Russet' apple, with blackcurrants and gooseberries hiding its rather garish looks.

I also have an equally wild allotment that is a tiny bit more intensive in production and has a small polytunnel. Between the two I have a year-round supply of fruit and vegetables meaning that the only fresh foods I ever need from the shop are carrots, leeks and onions, citrus and tropical fruit. Carrots elude me, perhaps because I was tormented as a child and called carrot top one too many times, and unless I grow them in pots, carrots either get munched by slugs or riddled with carrot fly. Life has many battles in it, but carrots aren't mine.

It is very hard to grow leeks and onions and not have them shredded by allium leaf miner in my part of the world. The same happens to garlic, but its tougher nature means that with a combination of my diversion tactics and its fortitude I can harvest enough garlic for the year. Rather than grow onions I grow perennial Japanese bunching onions which happily sit in a polyculture system and seem unbothered so far by the leaf miner. I tell you all this in part because every gardener, even those whose gardens look impeccable, has something that doesn't do well for them. By all means battle on till you master something, but life is also short. If your cabbage fails to head up and your Brussels always blow, choose something that won't vex you.

Finally, a note for gardeners who aren't in the northern hemisphere – from which this book has been written – to reverse any directional instructions, south-facing, north-facing and the like, to reflect your own whereabouts. In the northern hemisphere, south-facing walls and gardens are generally warmer and sunnier. In the southern hemisphere, the opposite.

What Is Polyculture?

At this point you might be wondering what a polyculture system is? The short answer is that it follows what nature does. It is a mixture, a little bit of everything, a muddle, a diversity of plants, a combination as opposed to a monoculture which is a single species usually grown at optimum spacing for speed of growth.

Polyculture can be wild and untamed, a space where you forage for food at the whim of nature's offering at that moment, or it can be very intensive and considered, a careful grouping of plants that will not inhibit each other's progress by competing for resources. In order for polycultures to work you must understand what happens both above and below ground. If you know what a plant's root mass looks like, then you've won half the battle in knowing who likes to be in bed with whom.

It is easy to imagine that all roots look more or else alike, a tangle of veins beneath the soil, but each genus of plants has wildly different patterns of growing. Some roots are storage organs such as tubers in potatoes, or root vegetables like parsnips and carrots and weeds such as dock plants, some others exploit the shallow surface of the soil like many annuals, and others bed down as deep as they can to wait the winter out. Then there are those that love to explore hidden areas, weaving their way through to new sources of food – climbers such as bindweed, blackberries, hops or wisteria.

There's another hidden layer that appears on all roots, the root hairs. Sometimes these are visible to the human eye; you can certainly see them on seedlings but when you pull up a beetroot you might imagine all the root mass is what's attached to the bottom of the vegetable, but there's a hidden network that spans far wider and deeper than you might ever imagine. Understanding what's happening below ground has made me a far more nuanced gardener than concentrating predominately on what's growing above.

Our desire to cram just a few more plants in small spaces is understandable. The consequenc-

es of this is often that root masses butt up to each other and this signals to the plant that there's not enough space to grow. In something like lettuce or beetroot this often results in early flowering or bolting as the plant decides to cut its losses and at least set some seed for future progeny, but in larger plants it's survival of the fittest and the one that can bully the most space for roots will win. This is the reason why you thin a line of seedlings. In a monoculture setting, ideal spacing for optimum growth ensures that root masses don't compete, but in polyculture it's a case of creating ideal bedfellows who don't compete and will help each other out. So when you look at a packet of seed and it gives you conventional spacing between plants and rows you have to work out the equidistance in either direction. Experimenting will teach you how far you can push this rule. For some vegetables I don't think in terms of distance, but of how much harvest I want. I can then work out how many plants I need and determine the spacing from that. In between the crops that sit in the bed the longest (parsnips, beetroot, sprouting broccoli, kales, cabbages and celery) I fit in the faster-growing stuff.

What Is Polyculture?

Why Bother?

It's entirely appropriate to ask the question why you'd go to the trouble of mixing everything up if ideal growth can be achieved by growing a single species. The answer is that nature doesn't like a monoculture, she likes diversity.

When a garden is healthy, abundant and productive it is diversity that makes it thrive. There are many players in the garden – beetles, worms, bees, millipedes, slugs, snails, nematodes, voles, moles, rabbits, squirrels, pigeons, songbirds, pets, lizards, mushrooms and on and on down to tiny things we can barely see with a naked eye and many, many more that we can't.

When a system is diverse no one player can dominate. If all you plant are potatoes and lettuces then the slugs are going to have a field day and even if the slugs don't the lettuce root aphids will, and the potato eelworms or some other pest because nothing is easier for a pest than giving it an all you can eat buffet of the same plants. If the pest does well then it can start breeding very fast. A single female aphid can give birth to 20,000 babies but their numbers are usually limited because there are plenty of others that like to eat aphids – parasitic wasps, hoverfly larvae, ladybird larvae, blue tits, wasps, spiders, fungi and bacteria to name a few.

Aphids are small and although they have a winged phase, they are at the whim of air currents and their tiny feet aren't going to march them far, so they rely on easy access, hitching a ride, hopping onto a neighbouring plant, being farmed by an ant. Life is easy for any pest if its nearest neighbour is an identical food source, not so easy if they're few and far between. Most pests are pretty specific to their hosts. There are hundreds of types of aphids – rosy apple aphid, black bean aphid, cherry black aphid, conifer aphid, pear bedstraw aphid, currant blister aphid, plum aphid, privet aphid, raspberry aphid, lettuce root aphid, woolly aphid and on and on. They may eat a few other plants, but they are surprisingly picky about their food. If you want to eat raspberry leaves and your next-door neighbour is a rose, or you want broad beans but you are sitting next to cabbage, you have to move a bit further to look for your supper. This not only uses resources but you might meet your maker on your way. The upshot is that you can't have an exponential population boom, you're kept in check.

Supermarkets have conditioned us to believe in the perfect-looking vegetable with flawless skin, of uniform size, unblemished and polished. Thousands are left rotting in the field because they don't conform, thousands more have vast amounts of chemicals drenched on them so that they can appear perfect. Perceived perfection is nearly always at the expense of flavour and if taste is prioritized then it's almost universally sweet these days.

Plant sugars make things taste great but we need vegetables also to taste of other flavours,

we need bitter, peppery, spicy, nutty, buttery, citrus-like flavours that talk of nutrients packed behind them. Many of us grew up with these supermarket vegetables and we may be going to the garden in search of wilder, deeper and more complex flavours but we take those ideals of what a vegetable should look like with us. Your vegetables may grow perfectly, but at some point they will look wonky, small, one end might be fat and the other thin, they may have holes, blemishes, rotten bits and dead bits and still taste delicious.

The brilliant food writer Jeffrey Steingarten wrote of a wonderful blind taste test experiment where lettuce and herb leaves with holes in them came out preferred to perfect ones. This is because the leaves with holes in them had more secondary metabolites in them. Secondary metabolites are the plant's pest defences, compounds that the pest doesn't like to consume. That holey leaf might look ridiculous but it also might taste of heaven.

When I feel defeated by slugs or battle-weary of flea beetles, I make myself walk through the garden and name all the things I can eat and use. Not the one failure that I'm hung up on, but all the other possibilities that might not at first glance seem like a whole lot, just a plant here or there, but put together they will always add up to a meal from my garden.

Often, a gardener may like to boast about how many pumpkins they have, the huge haul of onions or broad beans, but too often so much of those huge harvests go wasted. I have walked past allotments dripping in raspberries unpicked or courgettes turning into watery marrows that won't get eaten. Hopefully in each case the wildlife does well out of our excess, but all of this comes with other hidden costs. Think of the energy used to produce the seed or plants that were bought in, the resources in the soil to feed that plant, maybe chemicals to will them along, bamboo canes from faraway places, plastic netting that can't be recycled and on and on.

One of the joys of a polyculture mix that has a good backbone of perennial edibles is that even if the annual stuff evades you, the perennials will nearly always do fine. Their deep, established root systems mean that they have resources to weather a dry summer or a wet one, or a dry and wet one. They tend to do better with the vagaries of climate change and they will be the plants that bridge the gaps between crop failures and difficult weather. My perennial kales are always in leaf – whether I pick the most tender of the baby leaves to massage dressing into for a salad or shred and sauté the old leaves to bulk out supper, they are always there and I shall never garden without them.

Perennial vegetables and fruit also allow for adventures. I promised myself a long time ago I'd not chain myself to the garden, that I would go out and explore the world. I would never utter 'I can't leave the garden'. I fail on that last bit often, fretting that some baby plants won't survive not being watered every two seconds, but my perennial vegetables have taught me that I can go away and still come home to something for supper. They are the backbone that makes it such a healthy ecosystem, as the soil is not disturbed around their roots and this stability, plus their varied heights and sizes, make them particularly good homes for insects and other wildlife. Perennial fruit and vegetables are not resource-hungry either – I topdress them once a year with homemade compost and that is it. I water them when I first plant them to establish their root systems and do this regularly till I see signs of new growth, and after that it's up to them.

Every garden is different, what works in mine may not in yours. Observe closely, follow your hunches, but change variables one by one. It is tempting to want to throw everything at a problem, but then you run the risk of not finding out which method actually works. There's one exception though: when it comes to slugs, try everything – beer traps, midnight hunts, upturned empty grapefruit halves and cabbage leaves (they'll hide underneath them), ground eggshells and oyster shells, wool pellets, prayers, evocations and pleas!

How This Book Works

The premise is simple. Take a handful of plants from The Basics section (see page 25). These are all perennial and will make up the backbone of the garden – they give structure, year-round interest to your space, and in deep winter they might be all that is left standing. Choose wisely, you don't want to be moving this layer about every other year because you don't like the position. Be honest with yourself about how much space you have and what your conditions are truly like. Fruit trees are delightful but they won't grow in deep shade and nobody needs more than three perennial kales in their life.

Map out your basics onto a rough drawing of your space and see what space is left – now choose a smattering of fillers (see page 117). This section is made up of self-seeders and they will do their own thing and move where they wish. They're a playful lot and often weave the scheme together wonderfully, but you don't need many of them. However, if you are not going to be around for much of the time to do intensive gardening, this layer can be upped a little or a lot to give plenty of grazing and foraging with minimal effort.

The final layer is the intensive toppings (see page 133). These are your more traditional annual vegetables that often need to be raised indoors, in modules or seed trays or bought as young plug plants. They will need the most sun, the best soil and will require hours of your time to keep slugs off and flea beetles at bay. They will want feeding if grown in pots and some will need your warmest, most sheltered spot, even a greenhouse. They are a demanding lot, but the returns are utterly delicious. Some years my garden is made up of basics and a lot of toppings, other years it's the bubbles and froth of the fillers, the best years it's a good mixture of all three.

Before you start to choose your plants you need to contemplate your conditions and take into consideration how the light, in particular, works in your garden and how it affects who can sit where and why.

In ecology, a guild is a group of species that have similar requirements and play similar roles within the community and every plant in this book sits in a 'light guild'. There are those that tolerate deep shade, those that love dappled light shade that keeps their feet moist and those that must bask in the sun. Nearly all the toppings want full sun for instance, but many of the basics will tolerate quite a bit of shade.

Full Sun
Glade and Upper
Storey Plants

Those that like to bask in the sun fall into two very different groups. The tallest of these are the upper storey plants, which in this case are mostly fruit trees. Medlars and a few pears are the most shade tolerant of these, the others need as much sun as they can get to give ripe, heavy crops. At the other end of the spectrum are glade plants which are mostly herbaceous, and annual plants that have evolved either from the open glades found in woodland, when large mature trees topple and suddenly allow light in, or they've come from grassland, prairie, riverbanks, shorelines and meadows. Nearly all the toppings, the annual traditional vegetables, have evolved from grassland,

meadows, flood plains, riverbanks or shore edges.

If you are starting from scratch, you could use a sun-tracker on your phone or just go outside and observe the light conditions (taking into account how they will change with the seasons, with neighbours' trees coming into leaf) and map out which guilds you will need for your garden. This is particularly useful once you've shortlisted your upper storey trees and will give you a clear idea what spaces are left to fill. Then all that's left to do is not to get too carried away with eyes bigger than your belly or your plot. If you are adapting your existing space, take into account what is already there and fill in the gaps.

Deep Shade
Understorey Plants
and Climbers

Deep shade accommodates plants adapted to grow on the forest or woodland floor or to climb up into taller plants. They will receive no full sun and their leaves are adapted to eke out the most from the available light. They may be shiny to bounce light around or have deep purple or pale margins to absorb different parts of the light spectrum. They may also spend the summer dormant, making the most of spring light before the trees leaf out. Many spring bulbs fall into this category.

Light Shade
Lower Storey Plants
and Climbers

The plants growing in light shade are best thought of as the woodland-edge plants. They won't tolerate deep shade and they won't like full sun, but want a middle ground – a few hours of sun is generally fine as long as their roots don't dry out. In your garden many of these lower storey plants are shrubs and herbaceous plants that can be tucked into the edge of a tree's shade, in the shade of a fence or underplanted around roses and soft fruit bushes.

In a Nutshell

If you're overwhelmed by choice, or don't want to wade through a book, this section suggests a few reliable favourites for specific places.

The Potted Garden

There are many interesting ornamental edibles for a pretty, potted garden.

Bulbs (Glade Plants)

Mouse garlic, *Allium angulosum*
Nodding onion, *Allium cernuum*
Golden garlic, *Allium moly*
Welsh onion, *Allium fistulosum*
Chives, *Allium schoenoprasum*
Garlic chives, *Allium tuberosum*

Herbaceous (Lower Storey Plants)

Anise hyssop, *Agastache foeniculum*
Korean liquorice mint, *Agastache rugosa*
Campanulas, *C. portenschlagiana*,
C. poscharskyana, C. persicifolia
Miner's lettuce, *Claytonia perfoliata*
(let it self-seed in the shade of tall potted plants)
Alpine strawberry, *Fragaria vesca* 'Semperflorens'
Lemon balm, *Melissa officinalis*
Sweet violets, *Viola odorata* (let them self-seed in the shade of taller potted plants)

Trees and Shrubs (Upper and Lower Storey Plants)

Juneberries, *Amelanchier* species
Japanese quince, *Chaenomeles* species
Fucshia, *Fuchsia* species
Apple, *Malus domestica* on M27 dwarfing rootstock
Red and whitecurrants, gooseberries, *Ribes* species
Blueberries, *Vaccinium* species

Climbers

Blackberry, *Rubus* x hybrid
Japanese wineberry, *Rubus phoenicolasius*
Roses, *Rosa spp.*
Mashua, *Tropaeolum tuberosum*
Malabar spinach, *Basella alba* – not hardy, it must be brought in over the winter. This is not a spinach, but a tropical vine with heart-shaped, crisp, succulent leaves that can be used just like spinach. It is prolific in a sheltered spot (don't whatever you do put in a polytunnel or it will take over!) and can be grown in a pot. It needs something to climb up, such as pea netting.

Some Tips for Sowing and Planting Out Polycultures

There's a fine balance between sowing and planting densely with the intention to thin out (if the slugs don't), and over sowing. You need to get a good coverage quickly or else the weeds will get in, but you must thin. Remember that the thinnings are also supper. There's a window when the seedlings are getting their fourth, fifth and sixth true leaves where competition is everything – give them space and they will romp away, hold them back and the whole lot suffer.

Thin in succession; pull seedlings up roots and all to make room for the next wave rather than thinning to the final distance for a mature plant. Dominant, fast-growing plants in any polyculture quickly outshade the rest, but this can also be used to your advantage as they will hold shaded crops back and prevent them bolting – a trick that works particularly well if you've sown too many plants in modules and don't want to waste them. The shade of courgettes can be used to help blanch leeks and celery. Plant leeks in threes, celery alone between bush courgettes. Always have some backup Swiss chard; it is the one plant that doesn't seem to mind hanging out in a module or 9cm (3½in) pot and can be used to plug any unsightly gaps.

Some crops are just best planted in blocks: squashes will ramble over their neighbours too much, brassicas may need netting, potatoes are just terrible bedfellows. There are a few crops that work in between the bigger beasts: sunflowers and sweetcorn between squashes, lettuces under taller brassicas, quick crops of radishes before the potatoes exercise their land rights. Leeks don't like being thirsty so choose bedfellows that won't drain resources and that grow slowly – parsley, chervil and summer savory work well.

Start simple with three or four crops and don't try complicated polycultures of six or more crops, particularly if you don't have time to be harvesting often.

Remember that this is an intensive system and you have to repay your debt to the soil. Mulch with well-rotted homemade compost where you can to conserve water, keep down weeds and feed the soil. Use liquid feeds like comfrey tea and seaweed weekly if necessary, particularly when crops begin to flower and set fruit.

A dense covering of foliage means less water will be lost on the soil surface through evaporation, but in windy, hot weather the plants will demand more from the soil and may wilt. Don't water in the midday sun, it just creates demanding plants. A long soak in the evening is much better as the plant can take up the water slowly overnight.

Trees represent the backbone of the garden; they will give structure and interest all year round, offering dappled shade in summer, shelter to small things in winter and, if the earth is pleased, some very significant harvests. This lot are an investment; they may not look much in year one or two or even three for that matter, but they will mature to give your garden presence. They give the garden a range of heights which is not only aesthetically pleasing, it is hugely important for the wildlife. There is a direct correlation between varied habitats and diversity and many creatures don't want to live on the ground. A tree or mature shrub or bush offers a myriad of crannies and nooks.

Finally, this lot are very reliable. The slugs may eat your lettuces, the tomatoes may all get blight, but your perennial vegetables, herbaceous layer, climbers and fruit trees will weather any number of storms.

Their root systems are designed for the long haul, mining deep into the soil to get minerals so they have resources to take on a cool or hot summer or a hard winter. For this reason, you need to get this layer right, so take time selecting, talk to local nurseries about the best varieties for your location and make sure that the soil is in tip top condition when you establish them. The fillers and toppings in your garden can be changed on a yearly whim, but a fruit tree is for life.

Perhaps that well-known Christian origin story of the garden of Eden and its apple trees has swayed my psyche but a garden isn't a garden in my mind until it has a fruit tree in it. Its blushing blossom welcomes spring's true arrival, in summer it will cast a gentle dappled shade for those that burn easily, the many delightful woodlanders from wild strawberries to Solomon's seal, for it doesn't mind sharing its root space, and in autumn you get its crowning glory. There's a fruit tree for every size of garden from big old standard trees for a swing at the bottom of a longer garden to low-lying stepovers and wall-trained cordons for courtyards and patios.

The Basics
Trees

Apple

Malus domestica

POSITION *Sun to part shade* **SOIL CONDITIONS** *Fertile well-drained garden soil*
FLOWERING PERIOD *Early to late spring* **HARVESTING PERIOD** *Late summer to mid-autumn, storing into winter*
HEIGHT *Up to 9m (29½ft), dwarf varieties around 1.5–2m (5–6½ft)*
POLYCULTURE POSITION *Upper storey*

There are apple varieties that start ripening in mid-summer and others that are only just getting going in late autumn, and ones that only ripen in store so that it is possible to have fruit right the way through winter. I have ten apple trees between my two gardens, most as stepovers on the allotment. I don't get a huge harvest as stepovers, due to their size, are limited this way but I do get to gorge in early autumn and careful storage and drying means still have spring stores.

Apples are my true passion and I long ago gave up the idea that you should eat them all year round. I have no time for the bland supermarket apple that you can drop on the ground and it doesn't bruise. I'll hold out for my own apples that taste diverse with a hint of strawberries, or a touch of pineapple. Some are nutty and dense, some crisp and tart; they smell so heavenly that when I open the back door I can tell when they are ripe even all the way down at the bottom of the garden. These apples are rarely if ever in the shops, they hold a history in their DNA and I couldn't start to imagine a garden without a heritage apple tree in it.

Apples are truly giving trees, not least because you can fit one even in the smallest garden, thanks to dwarfing rootstock. And it's not just their blossom and harvest, the trees support so

much wildlife from nesting birds in older trees to bees feeding on the blossom and the many insects that find the cracks and fissures of the bark a good place to call home. Every year I find ladybirds tucked up asleep in my apples over winter.

The apple's origins lie along the valleys of the Tien Shan mountain range in central Asia and move along the Silk Road as apples fast became the snack food for weary travellers. Seeds and later cuttings were spread from traders to other parts of the world. Almost four thousand years ago the Mesopotamians stumbled upon grafting as a technique, further refining the selection of apples, which they passed onto the Persians and Greeks and then Romans. We have the last to thank for our apples, they brought us not just fine selected fruit, but the orchard economy too. Two thousand years and many tossed apple cores later there are around 7500 cultivars worldwide to choose from, 2000 in the UK alone.

In short, your perfect apple exists, you just might not have tasted it yet and it probably isn't for sale in the supermarket. I grow 'Katy', 'Discovery', 'Egremont Russet', 'Ribston Pippin', 'Ashmeads Kernel', 'Ellison Orange', 'Blenheim Orange', 'Sturmer Pippin' and 'Winston'. 'Discovery' and 'Egremont Russet' are grown as dwarf trees, 'Blenheim Orange' is a standard (the largest apple you can grow in my front garden) and the rest are stepover apples on the allotment.

Whatever apple you choose you do need to check whether it is self-fertile or if it needs to be cross pollinated with another apple tree to ensure good fruiting. Most are not self-fertile so they need a different apple to be in flower at the same time. It doesn't have to be the same variety, but it does need to be from the same pollination group. A bee will fly up to a mile or more for food, so the same tree doesn't necessarily have to be next door, but it does need to be within bee flying range.

If you are only going to grow one tree it has to be self-fertile, and it is always worth having at least one self-fertile tree even if you are growing a collection, as in a very wet spring it is often hard for bees to do their pollination work. 'Spartan', 'Red Windsor', 'Sunset', 'Scrumptious', 'Lord Lambourne', 'Red Falstaff', 'Laxton Epicure', 'Egremont Russet', 'James Grieve', 'Greensleeves', 'Braeburn' and 'Hereford Russet' are all self-fertile trees.

Apple trees are not traditionally grown with their own roots, instead they are grafted onto rootstock which determines their overall size from small, around 1.8m (6ft) to very large standard trees at 6m (19½ft) plus.

M27 is semi-dwarfing, tolerant of heavy soils, but doesn't like thin sandy soils and the tree will need watering, mulching and not too much competition from plants around it. It will also need permanent staking or the weight of the fruit will topple the tree eventually.

M9 is a commercial orchard rootstock, trees require permanent staking and don't like competition from neighbouring plants.

M26 is a general-purpose rootstock. You can use it to give vigour on poor soils for cordons and stepovers, it is tolerant of light sandy soils and suitable for container trees. Trees will need permanent staking.

M106 is a vigorous rootstock for trees up to

4–5m (13–16½ft); they tolerate light, sandy and wet soils.

M106 produces trees resistant to Phytophthora, woolly aphid and replant disease (where a tree is replanted in the same position as a related species). You won't need to stake these trees for life, but you may need a ladder to get the top apples.

M111 and M25 is a standard tree that will grow 6m (19½ft) plus in ten years. This is the traditional orchard apple tree with a straight trunk for space for a swing below. You will certainly need tall ladders and apple-picking baskets to get the harvest.

Pruning

You need to do some initial formative pruning over the first years and if this scares you, buy a pre-trained tree. These are certainly more expensive, but getting the shape right is half the battle and there are plenty of good nurseries that will help you get there.

After that annual pruning during winter is usual, when the sap has slowed down. Most apples are spur bearers, bearing fruit on short, fat spurs growing from the branch. Every year you should take 40–60 per cent of the previous year's overall growth (spurs, diseased, dead and dying material and anything that is rubbing or crossing) from the tree to stimulate new growth. However, a few apples ('Blenheim Orange', 'Bramley's Seedling', 'Discovery', 'Lord Lambourne' and 'Worcester Pearmain') are tip bearers, meaning the fruit is found on the tip of the previous year's growth, and if you prune this off, you lose the fruit. These trees are either left unpruned or pruned less frequently.

You don't have to prune at all; an apple left unpruned that is well fed and mulched produces exactly the same weight of fruit as a pruned tree, it's just the size of the apple that changes. Unpruned trees produce smaller fruit. If you are making juice or cider there's a strong argument that you don't need to bother with pruning.

I think of older trees, anything over 50 years old, as grandmas and on the whole they don't much like being pruned. They may not be that productive and will often fall into a biennial fruit cycle but this is the tree's way of dealing with pests; no fruit one year means you kill off the fruit pests. Pruning doesn't tend to stimulate more fruit production because they are at the end of their cycle. These trees may not give you buckets upon buckets of fruit but their wildlife value is huge, often housing many birds and insects, and they tend to look magnificent, all gnarled and holey.

Pests and Diseases

Unfortunately there's a whole suite of bugs hoping to claim your apples. Canker, *Nectria galligena,* is a fungus that causes lesions that look much like an open wound on stems and branches. It can cause serious damage to trees. Poor drainage and heavy soils exacerbate the problem. Affected branches should be cut out and disposed of.

Scab, *Venturia inequalis,* is another fungus that likes warm, moist conditions. It affects shoots, leaves and fruit with dull black or grey-brown lesions. Remove fallen leaves in the autumn as it overwinters in rotting leaves.

Codling moth is the bane of my trees. A common pest of apples, the larvae burrow into the young fruitlets and then proceed to eat them as they grow. They eventually crawl out to pupate within cocoons in cracks and fissures in the bark. Birds and bats are good predators, and you can use pheromone traps in spring to some success. I know one woman who puts pop socks over developing apples which deters the larvae, but that's some dedication on a big tree.

Bitter pit is so named because it causes brown spots with a slightly bitter flavour throughout the flesh, due to a shortage of calcium or water. The best control is through mulching and adding fast-acting lime or wood ash to the mulch in summer. This raises the pH of the soil, unlocking calcium to make it more readily available to the tree.

Crab Apple

Malus sylvestris

POSITION
Sun to part shade

SOIL CONDITIONS
Fertile well-drained garden soil

FLOWERING PERIOD
Early to late spring

HARVESTING PERIOD
Early autumn to late winter

HEIGHT
3–12m (10–39ft) depending on variety

POLYCULTURE POSITION
Upper storey

There are around 40 crab apple species, including the British native *Malus sylvestris*. The flavour and size of the apples vary greatly though all the crabs have small apples more suited to preserves than eating. Some make good small garden trees as they have upright habits, are self-fertile and flower for a longer period than apples. They can also be used as cross pollinators for apples.

'John Downie' has white flowers, orange-red fruit and particularly good autumn foliage. It grows to 10m (33ft) tall, but has a very upright and narrow habit so good for side returns and narrow gardens. 'Everest' is also upright, growing to 7m (23ft), with scarlet flower buds that open initially pink before turning white, then yellow-orange fruit that lasts well into winter before being devoured by the birds if you don't get there first. 'Butterball' makes a small tree at just 4m (13ft) with a spreading, almost weeping habit

with golden yellow fruit. This is a nice choice for a front garden as it elegantly fills your front window in summer, but lets light through in winter.

'Comtesse de Paris' is very like the popular 'Golden Hornet' but is a smaller, better-shaped and scab-resistant variety, to 4m (13ft), with butter yellow fruit that persist well into winter and make a lovely coloured jelly. 'Wisley Crab' reaches 4–6m (13–19½ft) tall and has very large fruit for a crab; rich purple-pink flowers are followed by purple-reddish fruit, like small apples with red flesh. They make a great juice but I think 'Harry Baker' is the king of crabs. Named after the superintendent of fruit at RHS Wisley, this tree has huge purple-pink flowers that open against purply foliage that gradually fades to green. Its fruit will win you prizes in the jelly and jam section of the local show. It's a big tree though; it often ends up 8–12m (26–39ft) tall.

Pears

Pyrus communis

POSITION
Sun, sheltered

SOIL CONDITIONS
Fertile deep well-drained soil

FLOWERING PERIOD
Mid- to late spring

HARVESTING PERIOD
Late summer to late autumn, storing into winter

HEIGHT
3–15m (10–49ft)

POLYCULTURE POSITION
Upper storey

Pear trees live much longer than apples, some up to 300 years, and if not grafted on to dwarfing rootstock they are taller, many growing to 15m (49ft). They need more sunshine and warmth than apples as they flower earlier in spring and are susceptible to late frosts. That said, it is very easy to wall or fence train them and they can even be grown as stepovers where space is tight as any pear is better than no pear to my mind.

There are around 3000 pear varieties worldwide and 300 or so are suited to the British Isles, though you wouldn't know this from the supermarkets that think there are just two, 'Conference' and 'Comice'. Both are fine pears, and worth growing, but you don't need to stick to them when you could choose 'Louise Bonne of Jersey' or 'Black Worcester' (mentioned in *A Winter's Tale* by Shakespeare), picked in late autumn and still perfect in storage in mid-spring. Find out from

your local fruit nursery what thrives in your area.

Pears do best trained on south- or west-facing walls or fences throughout the northern hemisphere, and don't mind a variety of soils as long as they never dry out in summer. Shade and competition from other trees will reduce their bounty. Most are grown on Quince A or Quince C and Pyrodwarf rootstock, the last two are best for wall-trained and dwarf trees which may need permanent staking. Nearly all pears are self-fertile so do not need another pear for cross pollination. They flower early, so pears largely rely on bluebottles to pollinate them (there's a reason not to squash them on the kitchen window), as well as hoverflies and wild bees.

All pears need to be pruned in winter, and wall-trained or stepover pears need to be pruned in summer as well to maintain shape and restrict growth. In winter you prune to thin the

spurs (the fruit buds) and occasionally remove a proportion of older branches to stimulate new growth. Formative pear pruning is very similar to apples but if in doubt invest in a pre-trained tree.

Unlike apples, do not let pears ripen on the tree. A pear that feels soft on the outside on the tree is often mealy on the inside. Also, the squirrels will have them, all of them, the minute you turn your back. To check whether a pear is ripe for picking, gently cup the fruit and lift and gently twist. If the pear parts easily from the spur then pick it and bring it indoors to ripen. Pears rarely all ripen together so check individually.

Like apples, scab and canker can both be a problem. To decrease susceptibility to scab make sure there's good circulation and shred or collect fallen leaves to prevent the cycle returning. For canker try and cut out all the infected wood before early winter. Recently European pear rust that causes bright orange splodges on the leaves and eventually canker-like swellings on the branches has been increasing. It looks very unsightly; you can prune out any cankers on branches, but it's best not to remove any infected leaves during the growing cycle as this may do more harm than good.

Trees

Medlars

Mespilus germanica

POSITION
Sun to part shade, any aspect, sheltered or exposed

SOIL CONDITIONS
Fertile, well-drained garden soil but tolerates heavy clay

FLOWERING PERIOD
Mid- to late spring

HARVESTING PERIOD
Mid- to late autumn to store for winter

HEIGHT
Up to 6m (19½ft)

POLYCULTURE POSITION
Upper storey

Medlars are strange and wonderful fruit that have been cultivated since the 13th century in the UK. There's something a little medieval about a fruit that you have to rot before you can eat it, spitting the seeds out as you do so. The correct term for ripening the fruit is bletting. It takes up to two weeks for this to happen after the fruit matures in early autumn. You pick the fruit when they are still hard and bring them indoors to blet somewhere cool. If you leave them to blet on the trees they tend to all drop and get stolen by the birds and squirrels that adore the flesh. Medlars taste rather like cooked apple and date with a swirl of custard and toffee. They can be eaten raw or made into fruit cheeses and jams and added to fruit leather mixes.

This tree is slow growing, up to 6m (19½ft) tall, and gently spreading, up to 8m (26ft) in width, with dark, twisted bark. Young branches are slightly downy, older branches can have spines and leaves are long, elliptical and downy on both sides. The flowers are quite beautiful – simple, rose-like, white turning pink as they fade. They open in late spring and are over by early summer. Fruits are apple-shaped, yellowish brown and downy; they are teasingly called arse fruit for good reason.

'Nottingham' is a UK cultivar that makes a loose-shaped tree with straggly growth, but tolerates heavy clay soils and crops well with good flavour. It will grow in the canopy shade of another deciduous tree so makes an ideal fruit tree if your neighbour's tree rather dominates your sun.

Quince

Cydonia oblonga

POSITION
Sun to part shade, south or west facing
SOIL CONDITIONS
Fertile, well-drained garden soil
FLOWERING PERIOD
Early to mid-spring
HARVESTING PERIOD
Mid-autumn, storing into winter
HEIGHT
3–5m (10–16½ft)
POLYCULTURE POSITION
Upper storey

Quinces are such beautiful trees, with pale pink cup-shaped flowers in late spring dancing on the end of elephant-grey branches and the bright, soft green of the new growth, to say nothing of the 'golden apple' that follows them. Quince fruit ripen late in autumn, headily fragrant, golden yellow with a soft down. All that perfume cooks to an aromatic, firm fruit that makes the best crumble, is divine stewed or baked into pie and ice cream, or creates the finest of confections, membrillo, to eat with manchego cheese. Quince were very popular during the 16th–18th centuries and there are numerous recipes for marmalades and puddings from that period, including very good wine.

Trees are grown on rootstocks Quince C or Quince A. Quince C makes a slightly more dwarf tree. Trees are self-fertile and generally reliable croppers. They can be trained as espaliers, fan trained or as loose bushes with multiple stems or a single trunk specimen. Quince need little pruning other than initial framework training and it is possible to buy pre-trained fan and espalier trees. Fruit is borne on spurs and tips of previous summer's growth so the initial pruning is carried out in winter, when leaders are cut back by a third of the season's growth to an outward bud. After that you don't have to do much pruning other than to remove dead wood. Freestanding trees may need staking for the first few years but after that they are fine, and tolerant of a wide range of soils, but very alkaline or chalky soils may cause yellowing of the leaves. They don't like cold winds, so need a slightly sheltered spot and some sun. Such a handsome tree, with little maintenance required.

Cathay Quince and Relatives

Chaenomeles cathayensis, Chaenomeles japonica, Chaenomeles speciosa

POSITION *Sun to deep shade, but fruits best in sun* **SOIL CONDITIONS** *Fertile well-drained garden soil* **FLOWERING PERIOD** *Early spring to early summer* **HARVESTING PERIOD** *Late summer to mid-autumn, storing into winter* **HEIGHT** *2–6m (6½–19½ft)* **POLYCULTURE POSITION** *Lower storey*

Usually known as Japanese quince, these small trees and shrubs are often found in urban front gardens as they are known to be extremely pollution tolerant, and had a bit of a moment in the 80s. *C. speciosa* and *C. japonica* are both shrubs that grow slowly to 2m (6½ft) tall. *C. cathayensis*, hailing from central China, is a tree that can grow up to 6m (19½ft), so you don't want to get them mixed up. *C. cathayensis* has the largest fruit, growing to the size of large oranges, whereas the other two bear satsuma-sized fruit on very spiny stems. The fruit are high in vitamin C and antioxidants.

These are undemanding types; sun is essential for good flowering, but other than that they will put up with a wide variety of conditions and do very well on clay soils. Once established they are fairly drought tolerant and require little or no pruning. They can be used in a mixed hedge or pruned into a hedge shape.

Loquat or Japanese Medlar

Eriobotrya japonica

POSITION *Sun to part shade* **SOIL CONDITIONS** *Fertile well-drained garden soil* **FLOWERING PERIOD** *Late autumn to early spring* **HARVESTING PERIOD** *Mid-summer* **HEIGHT** *Up to 9m (29½ft), but frost tends to keep it around 3m (10ft)* **POLYCULTURE POSITION** *Upper storey*

This small fruiting tree is an elegant evergreen for very sheltered gardens and courtyards. In truth it can grow huge, but tends to stop at around 3m (10ft) in milder climates. It flowers in late autumn to early spring so needs to be protected from very harsh frosts. Although well-established trees can survive down to -15°C (5°F), the flower buds are killed off at around -7°C (45°F).

Handsome large leaves grow up to 30cm (12in) long and 10cm (4in) wide, dark green and glossy above and covered in a rusty down beneath. The small orange fruit are borne in clusters, sometimes covered in down, ripening by mid-summer. Ripe flesh is succulent, tangy and sometimes very astringent. The fruit are quite moreish and can either be peeled and eaten raw or cooked into jams, jellies, chutneys and wines, but don't eat the huge brown seeds.

In a sheltered spot they are easy to grow and will tolerate some shade but grow best in full sun. They do need moist soil and dislike dry winds and dry conditions when the leaves will easily scorch. For the best fruit production a little pruning after harvest is required, reducing the terminal (topmost) shoots so that the tree doesn't grow too large, and keeping the centre open to allow light in. You can wall train plants into fans or espaliers which encourages prolific fruiting. This tree is an ideal subject for northwest-facing walls.

Fig

Ficus carica

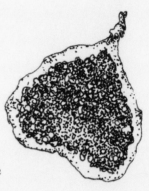

POSITION
Sun to part shade

SOIL CONDITIONS
*Any moisture-retentive soil type with
good drainage*

FLOWERING PERIOD
Early to mid-spring

HARVESTING PERIOD
*Mid-summer and again in early autumn
depending on variety*

HEIGHT
Keep restricted to 3m (10ft)

POLYCULTURE POSITION
*Upper storey, but can be trained to grow on
a wall or fence*

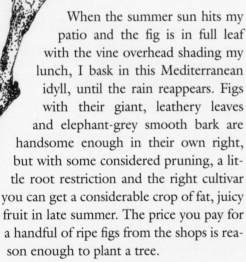

When the summer sun hits my patio and the fig is in full leaf with the vine overhead shading my lunch, I bask in this Mediterranean idyll, until the rain reappears. Figs with their giant, leathery leaves and elephant-grey smooth bark are handsome enough in their own right, but with some considered pruning, a little root restriction and the right cultivar you can get a considerable crop of fat, juicy fruit in late summer. The price you pay for a handful of ripe figs from the shops is reason enough to plant a tree.

Figs are self-fertile but they do need to have their roots restricted for productive growth; given free run you will get a huge tree – they can easily grow 10m (33ft) tall and spread half that – but little fruit of worth. The easiest way to do this is to plant the fig in a hole 60× 60×60cm (2× 2×2ft) lined with concrete slabs or bricks and packed with stones and broken bricks to 30cm

(12in) and then back filled with good garden soil and plant into that. Or you could plant into a similar-sized pot. I top dress with compost or seaweed once a year, but do little else.

'Brown Turkey' is the main fig growing in the UK. It is a seedless variety giving brownish purple fruit with rich, pink, sweet flesh early to mid-season from late summer into early autumn. It is good in pots whereas other available varieties, such as 'Mission', 'White Marseilles' and 'Desert King' need very sheltered, wall-trained spots to do well. A variety called 'White Icicle' has very distinct, snowflake-shaped leaves; it is mainly grown as an ornamental, but will fruit in hot summers.

Figs love the sun and need to bake to ripen their crop. They dislike cold and hate cold winds so may need protection, such as wrapping up branches with straw or bracken. Figs also hate having wet feet, and particularly hate cold, wet feet so you do need good drainage. In soils too rich and fertile you will get lots of growth and whopping great big leaves but almost no fruit, so you need to be mean to get your crop.

Figs look magnificent fan trained against a wall. They look triumphant and verdant in summer, and in winter their steely stems make quite a statement. You do need to be on top of pruning – in a mild climate figs produce two crops a year, but only one ripens. If you have hard, small figs on your tree by mid-autumn then you need to pick or knock them all off. They will not ripen and the tree uses up its resources trying to fatten a crop that will stay rock hard. The successful crop starts as embryo fruits (tiny fruits near the young shoot tips that are the size of large peas) in late summer and early autumn and these develop and ripen over the next summer into early autumn.

You prune to encourage as much young growth as possible. Prune in spring when threat of frost has passed, removing some older wood and shortening a few stems back to one bud in order to stimulate new growth. Any stems that are crossing, overcrowded or on a wall facing outward or toward the wall should be removed at their point of origin or back to a well-placed side shoot. Tie in any remaining shoots so they are evenly placed over the fan. In summer pinch out new shoot tips to leave five to six leaves. In pots, every spring cut away three or four of the oldest or weakest stems to the base to give room for new growth and in summer prune new shoot tips as above.

Plums and Gages

Prunus domestica

POSITION
Sun, must be out of frost pocket or windy site for pollination, best in south- and west-facing positions, but there are varieties suitable for north-facing sites

SOIL CONDITIONS
Highly fertile, moist but never waterlogged clay or loam

FLOWERING PERIOD
From early spring, flowers are often killed by frost

HARVESTING PERIOD
Late summer to early autumn

HEIGHT
2–6m (6½–19½ft)

POLYCULTURE POSITION
Upper storey

Plums are prolific, branches drip with juicy fruit even on young trees. For the first few days this feels like you've won the fruit lottery, but plums don't store well, so very quickly you have to move into serious preserving – jam, ice cream, wine, dried, stewed, frozen, fermented and then you're round to the neighbours with bowls full.

A good harvest comes from a happy tree and plums need sunny, sheltered sites free from late spring frosts as plums flower in mid-spring when they are just coming into leaf. They like slightly acid soil; clay is good, but wet, badly drained or sandy soils aren't. Mulch every spring with garden compost, which will prevent drought stress in early summer, keep down the weeds and feed the tree – extra nitrogen such as a feed of chicken manure pellets in spring is ideal.

Plums are a diverse bunch and may be self-sterile, partly self-fertile or fully self-fertile. Clearly the former two will need a pollination partner to get things going. Your local fruit tree nursery will know the details, make sure to ask. Formative pruning is very important because those heavy crops will break branches if a good framework isn't established. Young trees are pruned in early spring and older trees in midsummer. Plums must never be pruned in winter, this causes silver leaf which can lead to bad canker, a fungus that enters the tree through a fresh wound and can eventually kill the whole tree.

Pruning is quite basic: anything crossing, weak, badly placed, diseased or dead is removed. Well-placed growth can be left unpruned, but branch leaders that are weak or growing horizontally are pruned back by a quarter in early spring to early summer to an outward facing bud.

There are hundreds of varieties of plum that are broadly broken down into dessert plums (too low in acid for cooking), cooking plums and greengages. Try and find a variety that is local to your area. Dessert plums, such as 'Blue Tit' and 'Opal' are the finest, but come with all sorts of high-maintenance requirements, needing warm, sheltered, very sunny spots. Cooking plums are more acid, but more shade tolerant. Varieties like 'Czar' and 'Pershore' are also very good eaters and can be grown on north-facing walls, making them the ones to go for if you're limited with space.

Greengages are halfway between a dessert plum and a damson, often green or yellow and the stuff of childhood dreams, juicy, sweet with a distinct rich flavour. 'Cambridge Gage' and 'Imperial Gage' are both reliable crops from late summer onward. The trees are less vigorous – a healthy plum can grow to 9m (29½ft) – but they absolutely require sheltered warm spots.

There are numerous rootstocks for smaller gardens; semi-dwarfing are best, producing trees that grow 3–4m (10–13ft) high with good yields. Pixy rootstock is semi-dwarfing and best suited for cordons, fan-trained and bush trees. VVA-1 is better for harsher winters. St Julian A is semi-vigorous, making a tree that is 4.5–5m (about 15ft) tall and suited to a wide range of soil conditions.

Wasps can be a pain with ripe fruit; they can seriously damage the crop and you as you try and fight them for it. Sometimes it may be necessary just to pick the crop early – plums ripen very well indoors. Or make a pond – dragonflies are serious predators of wasps.

Juneberry or Serviceberry

Amelanchier lamarckii

POSITION
Sun to light shade
SOIL CONDITIONS
Unfussy, will tolerate heavy clay
FLOWERING PERIOD
Mid- to late spring
HARVESTING PERIOD
Early to mid-summer
HEIGHT
2–6m (6½–19½ft)
POLYCULTURE POSITION
Upper storey

Amelanchier lamarckii is perhaps the easiest of the Juneberries to get hold of and can be grown as a shrub, making it perfect for smaller gardens, or it can even be grown as a hedge. The fruit appears in early summer and by mid-summer a tree can be covered in tiny, purple-blue berries that taste remarkably like blueberries with a strong hit of apple. The blackbirds in particular all seem to know how good these berries are so you need to keep an eye on how they are ripening to get in quick for your share.

I pick and dry the fruit which I add to muesli and porridge through the winter. Rehyrated seeds work well in muffins and cakes too. You can either dry the fruit in a dehydrator or on baking trays out of direct light. The fruit is rich in iron and copper and the seed has a slightly almond-like flavour.

These are the greens that will feed you when you've been absent from the garden, the ones that will put supper on the table when the slugs have got everything else, that grow either despite the lack of sun or because there is too much. Many also reliably create a strong, even architectural backdrop for the summer-flowering froth and annual vegetables.

The Basics
Leafy Greens

Kale, Bush Kales, Cottagers' Kales

Brassica oleracea var. *ramosa*

POSITION
Sun to part shade

SOIL CONDITIONS
Any good well-drained garden soil

FLOWERING PERIOD
Late spring to early summer

HARVESTING PERIOD
Year round

HEIGHT
40cm–1.2m (16in–3ft 10in)

POLYCULTURE POSITION
Glade

This is one of the earliest cultivated forms of brassica, known to have been grown since Roman times. Some forms still grow wild on British cliffs today. The forerunner to the cabbage grows as an evergreen bush, often quite sprawling, usually multi-stemmed and flowering infrequently. These are not neat plants, tending to flop to the ground where their trailing stems root before ascending again.

The most readily available variety is known as 'Daubenton's Kale', named after the 18th-century French naturalist Louis-Jean-Marie d'Aubenton who found it growing in eastern France. It is a multi-branching bush kale with leaves up to 15cm (6in) long and good kale flavour. I like the variegated form with its white edges and slightly bluer green foliage because it looks so good in spring with tulips, forget-me-nots and honesty.

These kales do best in full sun, in rich, heavy, slightly acidic soil, but are tolerant of a wide range of conditions though they can be seriously attacked by slugs in shade. In dry periods, cabbage aphids have been an issue and any pigeon will make mincemeat of the plant in winter if it can, particularly if it can perch on the plant or nearby.

Dense planting of other things around the base, for instance winter seed heads of honesty, can act as a foil against pigeons. Cabbage white butterflies can be an issue in summer, but these kales are much tougher than most cultivated brassicas.

Plants should be propagated every three to four years to maintain vigour. Do this by taking cuttings in spring and summer. Take a leafy side shoot about 15–20cm (6–8in) long, remove the lowest set of leaves so that you have 5–8cm (2–3in) of stem and insert the cutting into a pot of moist compost. Keep this somewhere out of direct sunlight. You can cover the cutting with a ventilated clear plastic bag (poke a few holes in it) to keep humidity up. Don't worry if the upper leaves fall off, this is normal, and be patient as the cutting can take over two months to root. When you see new roots emerge out of the base of the pot and new growth, the cutting is ready to plant out, taking precautions against slugs.

In theory you can eat any leaf, even the oldest, but oh boy are they tough, so in reality go for the young, tender leaves toward the top. You can harvest whole side branches too, as this helps to keep the plant in shape.

Tree Cabbage, Jersey Cabbage or Walking Stick Kale

Brassica oleracea Acephala Group and *Brassica oleracea* var. *longata*

POSITION *Sun to part shade* **SOIL CONDITIONS** *Any good well-drained garden soil*
FLOWERING PERIOD *Late spring to early summer* **HARVESTING PERIOD** *Year round*
HEIGHT *1.5–2m (5–6½ft)* **POLYCULTURE POSITION** *Glade*

These wonderful tall cabbages hail from Europe, mainly from northern Spain and Portugal, producing one large head of kale on a single long stem. They will live for two to three years and if you keep removing the lower palm-shaped leaves to encourage the stem to grow you can polish and lacquer it into a walking stick if you so wish.

My favourite is 'Paul and Becky's Asturian Tree Cabbage' from Asturias, with wonderfully flavoursome lemon green leaves. You can keep pinching off the flower stalk so it continues to produce leaves. Eventually I let it flower and self-seed and this way I've had continuous production of the plants for years. I rehouse seedlings wherever they spring up on the allotment into the border at home as this plant looks so good backlit with the late summer sun. One joy of tree cabbages is that their long, tall stems allow plenty of other plants to grow underneath them, anything that doesn't mind a bit of dappled shade will thrive.

'Nine Star' Perennial Broccoli

Brassica oleracea Botrytis Group

POSITION *Sun to part shade* **SOIL CONDITIONS** *Any good well-drained garden soil*
FLOWERING PERIOD *Late spring to early summer* **HARVESTING PERIOD** *Early to mid-spring*
HEIGHT *40cm–1m (16in–3ft 2in) tall* **POLYCULTURE POSITION** *Glade*

This is a cauliflower masquerading as a sprouting broccoli that doesn't want to die. 'Nine Star' is the only variety in cultivation that I know of. The name comes from its growth habit, with one big central flowerhead around which up to nine smaller heads appear. It grows to 60–90cm (2–3ft) tall and wide and tends to flop a bit. It is a short-lived perennial; if you're lucky you can get four years before it tends to start splitting at the base but it is easily propagated in spring from side shoots, in the same way as perennial kale, or you can grow it from seed.

Harvest the main flowerhead as soon as it is big enough to be worth eating as this encourages side shoot formation. You can usually harvest heads for a couple of weeks and I grow two plants to get a decent amount and continue to harvest young, small leaves through the summer. As plants sprawl with rather ugly legs it's good to underplant – I use Canadian honewort, *Cryptotaenia canadensis,* that looks very pretty when its flowers appear between the leaves.

'Nine Star' likes good fertile soil in sun or light shade, and the usual brassica pests of slugs, caterpillars and aphids can all be a problem, as well as pigeons. If damage gets too bad, coppice the whole plant back to a node on a stump 15–30cm (6–12in) tall and it will usually resprout.

Perennial Wall Rocket or Wild Rocket

Diplotaxis tenuifolia

POSITION *Sun to part shade* **SOIL CONDITIONS** *Any good well-drained garden soil*
FLOWERING PERIOD *Mid-summer* **HARVESTING PERIOD** *Year round*
HEIGHT *20–45cm (8–18in)* **POLYCULTURE POSITION** *Glade*

This is a small, but very useful green for the front of the border or bed. Perennial rocket is much like its annual cousin, but with a skinnier leaf and slightly more peppery taste. It has very pretty bright yellow flowers on long, wiry stems and will happily self-seed, so once you have it in the garden you don't have to propagate it, just allow it suitable living space. On rich soil it is not very long-lived.

As the name suggests it likes free-draining cracks and nooks, walls and rock faces and will self-seed itself into improbable places – flavour can get very peppery on poor soil. Leaves and flowers can sometimes be harvested year round as it will still produce greens in a mild winter. Flea beetle may damage the leaves, but it will outgrow the problem.

Sorrels

Rumex species

POSITION *Sun to part shade* **SOIL CONDITIONS** *Any good well-drained garden soil*
FLOWERING PERIOD *Late spring to early summer* **HARVESTING PERIOD** *Year round*
HEIGHT *10–60cm (4in–2ft)* **POLYCULTURE POSITION** *Glade*

There are numerous sorrels with delicious tart leaves that provide a welcome green in late winter and early spring when little else is up. Garden sorrel, *Rumex acetosa,* is a deep-rooted perennial that will live for several years before it sends up a flowering shoot, then the plant's leaf production will start to decline. 'Profusion' is a non-flowering variety that is very useful for this reason, but you can also just leave sorrel to flower and it will merrily self-seed. 'Blonde de Lyon' and 'Large de Belleville' are French broadleaf varieties that are well worth seeking out.

Sorrels are deep rooted and rich in potassium and phosphorus. Most are herbaceous, dying back in deep winter, but they reappear early. Sorrel is delicious raw used sparingly in salads, or in salsa verdes. If you want to cook with it, it is particularly good blended into a sauce or added into creamy mash potatoes. Cook it for no more than 60 seconds in boiling water so that it retains its colour and flavour. If you overcook, it goes a particularly unappealing brown.

Scorzonera

Scorzonera hispanica

POSITION
Sun

SOIL CONDITIONS
Any good well-drained garden soil

FLOWERING PERIOD
Mid-summer

HARVESTING PERIOD
*Roots are harvested from autumn to spring,
young shoots in early spring and unopened
flower buds in early summer*

HEIGHT
2.2m (7ft) tall in flower

POLYCULTURE POSITION
Glade

I grow a lot of scorzonera on the allotment. It's usually sold as a heritage root vegetable but although the roots are tasty they are quite a pain to process and clean, whereas the leafy parts are a doddle. Scorzonera is essentially a huge dandelion with flower stalks that can easily reach 1m (3ft 2in) tall. Flowers are bright yellow and the tap root is long and black with white flesh. Closely related is salsify, *Tragopogon porrifolius,* which has white roots. Salsify has a lovely purple flower and will happily self-seed, but unlike scorzonera, is needs to be eaten before it flowers as it's biennial.

Young leaves can be eaten raw when they have a nice crunch and are similar to lettuce, or older cooked like spinach or endive. Unopened flowers heads are delicious, I eat them fried with garlic on pasta, and raw. The roots bleed easily and need to be peeled before you eat them, which is easiest done after they're cooked. They can be boiled or steamed and have a mild, pleasant flavour.

Scorzonera is sown in spring, directly or in modules. If sowing direct sow in a line and transplant young plants 10cm (4in) or so tall to their end position. Roots are harvested in autumn and winter and get considerably bigger from year two onward; the quality doesn't decline with age. The roots snap easily so dig with care. For best root production you need to grow in quite sandy conditions, but if you're not after fat roots, then the plant is unfussy as long as it gets some sun and the conditions are not waterlogged.

Nettles

Urtica diocia and relatives

POSITION *Deep shade, part shade and light sun* **SOIL CONDITIONS** *Good garden soil, fertile, well drained*
FLOWERING PERIOD *Early to late spring* **HARVESTING PERIOD** *Leaves are mostly harvested during late*
winter and spring **HEIGHT** *10–30cm (4–12in) when in flower* **POLYCULTURE POSITION** *Glade*

If your garden is big enough there are plenty of reasons to allow a few wild margins to entertain a little stinging nettle action. It has numerous uses including as a delicious and nutritious edible spring green, and can also be rotted down as plant food. I understand that you may not want to actively make space for a rampant weed, but two very beautiful and useful nettle relatives are definitely worth considering, both from Japan.

Boehmeria tricuspis var. *unicuspis* has won a place in my heart for its handsome autumn looks. Resembling a stingless nettle, by autumn its stems have flushed red and it's covered in long, pipecleaner-like pinkish flowers and seeds. It doesn't spread, loves shade to partial sun and grows to 60cm (2ft) tall. The young leaves are edible, cooked just like stinging nettles or spinach. *Elatostema umbellatum* 'Dents de Kyoto' is a low-growing, 20–30cm (8–12in) tall nettle relative with arching stems of pointed, highly toothed bright lime green leaves, and small clusters of white flowers that eventually turn to purple bulbils. The young foliage and stems have a fresh floral taste, slightly sweeter than our nettles. It loves deep damp shade and dies back in winter. Harvest once your clump has become established.

Soloman's Seal

Polygonatum species

POSITION *Part shade* **SOIL CONDITIONS** *Any good well-drained garden soil*
FLOWERING PERIOD *Late spring to early summer* **HARVESTING PERIOD** *Early spring*
HEIGHT *60cm–2m (2–6½ft)* **POLYCULTURE POSITION** *Understorey*

My plants grow under my apple tree and every spring I am surprised and delighted by their return. The young unfurling shoots can be harvested just like asparagus and taste so similar it is a wonder that this is not a more popular vegetable. It's certainly far easier to grow. A woodland perennial, plant in part or total shade or part sun, and wait for your plants to form a thicket. Once this is large enough, which takes around two years, you can start to harvest. Solomon seal sawfly larvae eventually rip the leaves to tatters in my garden but not until after I've harvested what I want.

Harvest young stems when 20–30cm (8–12in) tall before the leaves have fully unfurled. The stem tastes of sweet asparagus. Only take one cut in spring as the plant needs to recover. Solomon's seal is tricky from seed so buy plants. Edible species include the garden Solomon seal, *Polygonatum x hybridum*, common Solomon's seal, *P. multiflorum*, angular Solomon's seal, *P. odoratum*, and whorled Solomon's seal *P. verticullatum*. They all reach about 60cm (2ft) but giant Solomon's seal, *P. commutatum*, is a bit of beast growing to 2m (6½ft).

Cuckoo Flowers

Cardamine species

POSITION *Deep shade, part shade and light sun* **SOIL CONDITIONS** *Good garden soil, fertile, well drained*
FLOWERING PERIOD *Early to late spring* **HARVESTING PERIOD** *Leaves are mostly harvested during late winter and spring* **HEIGHT** *10–30cm (4–12in) when in flower* **POLYCULTURE POSITION** *Glade*

A number of UK native cuckoo flowers are worth foraging for. The tiny delightful pale pink flowers and young leaves of *Cardamine pratensis* can be picked in early spring for a sweet peppery hit, or in winter wavy bittercress, *Cardamine flexuosa*, can be weeded from around pots and disturbed ground for a watercress-flavoured hit in salads. These two grow wild, but *C. pratensis* 'Flore Pleno Alba' is worth looking out for, and I have become fond of the greater cuckoo flower, *Cardamine*

raphanifolia. Spreading out to 60cm (2ft) and up to 40cm (16in) tall, this has large lilac-pink flowers that are an important caterpillar food for the orange tip butterfly. Its large, watercress-flavoured leaves are abundant all winter long. By summer the whole plant tends to die back to its rhizomes in my garden but it doesn't mind being shaded out by my other greens and vegetables. It thrives in both shade and part sun and prefers damp ground, so could be encouraged around a pond.

The shrub layer of the garden is vital, but easily overlooked, particularly when space is at a premium. Carefully placed shrubs hold the garden together, can be used to hide unsightly bits, and will create different heights and niche micro environments. And perhaps more importantly they create the necessary layers for a thriving ecosystem. Different folk live in different places – shrubs create those niches from hidden understoreys to dense branches that act as habitats for insects, cover for birds and nesting sites.

The best shrubs double up as food for us and for wildlife but I would caution anyone against hasty removal of existing shrubs just because they aren't food. They are important habitats, so while you get the rest of the garden going perhaps they can stay? If they are too large, can you prune them back? Can they be used as structure to grow edibles up or through? The wilder tomatoes, like 'Matt's Wild Tomato' will scramble through rose buses and shrubs happily if planted in a sunny spot next door. Climbing squashes and achocas (cucumber relatives) will cheerfully clamber up and over them, and Jerusalem artichokes will muscle in around the base of shrubs and grow up and through them using the shrubs as structural support.

Without doubt the best edible shrubs for the natural garden are soft fruit. I've never covered any of the soft fruit in my garden with netting. I find if you weave them into the scheme with a perennial and annual understorey they are hidden enough from the birds and I'm happy to share the end of the harvest with the odd bold blackbird or songbird, it's as much their garden as mine.

The Basics
Shrubs

Currants

Ribes species

POSITION
Sunny, light shade

SOIL CONDITIONS
Tolerant but prefer well-drained, moisture retentive soil

FLOWERING PERIOD
Early to mid-spring

HARVESTING PERIOD
Early to late summer

HEIGHT
60cm–1.8m (2–6ft)

POLYCULTURE POSITION
Lower storey, glade

White are certainly more beautiful, red are more prolific, but black are everything. Compared to the other two the berries are far sweeter and, thanks to the blackcurrant juice drink Ribena, they have had far more breeding. There are now very easy-going bushes that keep a nice tidy shape, don't grow too wild or unwieldy (unlike red) and are packed with big, fat, juicy berries, as long as they get some sun.

Still, if you have space for all three, I'd make a claim for doing just that so you can make a champion of summer puddings. White and redcurrants are not sweet; they need a ton of sugar to make them palatable, but once that's sorted, they are lovely. Redcurrants are always overly prolific and most people who grow them find they still have a freezer full of berries when they come to add the current year's crop. Whitecurrants are perhaps a little more useful as they will grow in shade. This does nothing to add any sweetness, shade actually makes them tarter, but out of the three, whitecurrants do possess the strongest essence of currant. They taste quite sublime (if tart) and they do look like jewels on your cereal, pastries, desserts or frozen for cocktails.

When you first plant a blackcurrant you have to be brutal and cut the whole plant back to about 2.5–5cm (1–2in) above the soil level (you can take cuttings as this point). I know this means you won't get any fruit this year, but if you fail to do this the plant doesn't send enough energy down to its roots and every season afterward produces a slightly lacklustre crop.

Blackcurrants crop on last year's young growth, so pruning is all about stimulating new growth. Prune in winter by taking out up to a third of the oldest growth. Blackcurrants root incredibly easily, so take a pencil-sized new growth about the length of your secateurs, plant this in a pot and the cutting will have rooted by spring. Varieties are broken down into early, middle and late season, but as this is all happening in a matter of weeks in mid-summer, I wouldn't worry too much about which one you go for as long as you get a compact form. Some commercial beasts like 'Boskoop Giant' would swamp a small garden.

Early season 'Ben Connan' has large berries and is a heavy cropper but still compact enough for smaller gardens. 'Ben Sarek' is perhaps the best for a small garden; some say the berries are

slightly acidic but as you're inevitably going to add sugar anyway, that's not a great issue. In a heavy-cropping year the branches will sprawl so it needs some staking. 'Ben Tirran' has very good flavour and large berries that crop late, a good choice if you have space for more than one. It's a vigorous grower, but stays fairly upright, so works well with other perennials around it.

You can grow all currants as bushes, but if you want to fit all three in try growing the red and white as upright cordons against a fence or wall. They are not fussy about soils and both will tolerate partial shade. You may need to go to a specialist fruit nursery for cordons as most currants are sold as bushes, and buy them barerooted in winter, when they'll cost a fraction of price of a container-bought plant in spring. Cordons can be planted 30cm (12in) apart, so you can get a lot of different varieties into a small space. You can also create a double cordon, where you have one single stem that branches into a U shape, a highly attractive system along a wall. Red and whitecurrants will both grow on northwest-facing walls. Plants need to be well mulched every spring.

To prune a red or whitecurrant as a cordon,

cut back the tip of the main growing stem by a quarter to just above a bud on planting then remove all side shoots that are 15cm (6in) from the ground or below and cut back any higher side shoots to one or two buds. The next year from early to mid-summer cut all young shoots to five leaves and tie in the growing tip to a cane to keep it upright. In late autumn, after leaf fall, prune back the same side shoots to one or two buds and cut back the tip by a third. Eventually your cordon will get too tall for its support, then cut back the tip to five leaves from last year's growth in summer, and in winter reduce this to one to three buds.

If you want a bush, in the first year of planting in spring choose four or five well-placed stems and cut these back to 15–20cm (6–8in) to create a framework. Remove all other stems back to the ground. In year two and onward, in early to mid-summer shorten all sideshoots to five leaves and in winter shorten all tips by a third and the side shoots back to three buds.

Or don't bother. The truth of the matter is that if you just prune out in winter anything that is diseased, dead or dying, or anything crossing, and shorten the tips by a third, the currants will be fine.

Gooseberry

Ribes species

POSITION
Sun to part shade

SOIL CONDITIONS
Good garden soil, well-drained, avoid water-logged soil

FLOWERING PERIOD
Early to mid-spring

HARVESTING PERIOD
Early summer to early autumn

HEIGHT
60cm–1.8m (2–6ft)

Gooseberries are treated just like red and white-currants. They make fine cordons or large, prickly bushes and they always produce plenty of fruit. If growing as cordons, thin the fruit in late spring and early summer. If you take out every other fruit, the gooseberries grow big and fat and certain varieties of dessert gooseberries turn into the finest fruit you may ever taste. There are hundreds of varieties, coming in purple, pinks, reds and yellows as well as green.

I have in the past been very dedicated to gooseberries and grown rows upon rows of cordons, but life is short and there is much to see in the world. These days I have three bushes, two are the modern Finnish-bred, disease-resistant 'Hinnonmaki Red' and 'Hinnonmaki Green', which are sweet dessert types and if I am very honest, I don't even prune them.

The green one grows underneath an apple or perhaps is better described as growing through the apple. It does whatever it pleases and gives me more fruit that I know what to do with.

The red one lives on the allotment along with 'Invicta', a cooker/dessert variety, meaning it can be picked young to cook with and keeps its shape, or thinned to become a very tasty dessert gooseberry. Both are grown as bushes, pruned to be a little neater than the one under the apple, but only just. They all give phenomenal harvests.

Raspberry

Rubus idaeus

POSITION *Sun to part shade* **SOIL CONDITIONS** *Good garden soil, prefer slightly acidic, well-drained, won't do well in shallow chalky soil or waterlogged conditions*
FLOWERING PERIOD *Early to late spring* **HARVESTING PERIOD** *Early summer to mid-autumn*
HEIGHT *1.5–2m (5–6½ft)* **POLYCULTURE POSITION** *Lower storey, glade*

The joy of raspberries is their ease as long as you recognize one thing: they are restless sorts that like to move about. Raspberries are shallow rooting, which means that their roots are easily disturbed by weeding and hoeing and that they use up the food source of this top layer very quickly. All that towering growth in a year means they are hungry feeders.

You can tell when your raspberries are bored with their spot because, whether you like it or not, they will start moving – their underground runners will pop up wherever they think is best, sometimes surprising distances from their mother plant. If you force them to stay put by endlessly removing any suckers that appear elsewhere, you'll find they sulk. Often the leaves will start to yellow between the veins, a sure sign they are running out of magnesium.

In a small garden, it is not always possible to let them roam as they wish, so get round this by replanting healthy suckers when the plant is dormant in winter. You can also divide large clumps to rejuvenate vigour. Once you have raspberries that you like, use this technique and you should never run out of new plants and have plenty to give away.

For the best berries mulch twice a year with homemade compost, in early autumn and again in spring. The autumn mulch will protect the roots from too much winter wet while rotting into the soil in time for spring growth. The spring mulch will keep down weeds and lock in moisture – if mid-spring is dry this tends to effect flower formation – and give food for summer growth.

Raspberries come in two forms, autumn (primocane) and summer (floricane), and each category has a very specific pruning regime. Autumn canes fruit in late summer to early autumn and do so on this year's canes. They are pruned in late winter, cutting all canes right back to ground level to make space for the new lot of canes.

Summer raspberries fruit mainly in mid- and late summer, though there are some very early and very late cultivars either side of that. They fruit on second year growth so if you prune them like autumn ones, you'll not see a single raspberry ever again. Instead in summer, once the fruit has finished, you cut the spent canes right down to the ground, leaving no stub. The new growth for next year is already up and growing. The simplest way to tell the difference between the spent canes and the new ones is colour, new canes are reddish brown while the spent canes are bleached tan and woody looking. You should thin the new growth so that there are no more than eight or so stems per plant. Any weak, thin or badly placed new canes should be removed to ground level.

Traditionally you would then tie in all the new canes to wires and posts, but this is not only regimented and rather ugly looking, but stupid (I've finally said it). Raspberries naturally protect their fruit from the birds by bending over like a walking stick, the fruit being nestled between the leaves and swaying so that only the lightest of birds can land on them. This preserves the fruit until it is ripe (and thus the seeds developed) enough to fall off and become everyone's game. If you tie them into posts and wires you expose all this fruit and give the birds something to sit on while they harvest, which means you have to make a fruit cage and cover them. It makes sense for commercial picking to preserve the fruit (though most these days are grown in tunnels), but for those of us with wild back gardens letting the raspberries sway gracefully is a far more elegant solution.

There are many varieties to choose from. 'All Gold', a delightful yellow variety, and 'Autumn Bliss' are the standard for autumn canes. For summer 'Ruby Beauty' is an early fruiting compact form that's good for small spaces and container growing. 'Malling Jewel' is another compact form ready mid-season. 'Glen Ample' produces an obscene amount of fruit on spineless canes so is particularly good when small people are harvesting. 'Leo' is a bit of beast, with long, vigorous canes, but satisfyingly firm fruit with good flavour appear toward the end of summer, just before the autumn canes set off.

Blueberry

Vaccinium corymbosum

POSITION
Sun to very light shade

SOIL CONDITIONS
Acidic soils (pH5.5 or lower), free draining, moisture retentive

FLOWERING PERIOD
Mid-spring to early summer

HARVESTING PERIOD
Mid-summer to early autumn, storing into winter

HEIGHT
50cm–1.8m (20in–6ft)

POLYCULTURE POSITION
Glade

Blueberries are truly the luxury goods of the fruit world. They are famously acid loving meaning that they must be grown in ericaceous compost that has a suitable pH or else they will sulk. Usually this means they are grown in pots where they do very well as long as you keep on top of watering and repot, preferably every year, as they are hungry feeders. Ericaceous compost tends to tire quickly and compacts, which causes the blueberry roots to rot. You can repot into the existing pot if necessary. I tend to do this in late winter or very early spring before the plant is in leaf. If you can't repot, at least top dress with new compost.

The other alternative is to grow the blueberries in the ground. You can do this even if you don't have acid soil by digging ericaceous compost into the planting hole, preferably in autumn, to plant in spring. Then regularly top dress with ericaceous compost as mulch and use pine needles (recycling the Christmas tree is ideal) or pine bark as mulch. Don't use manure or mushroom compost as these tend to be quite alkaline.

In my experience blueberries take quite some time to settle into new soil, but after two years or so they don't much seem to mind that they are not in acid soils and produce healthier, plumper fruit when grown in the ground. The other thing to note is that blueberries do much better in pairs or more, cross pollination improves fruiting no end.

There are three types of blueberries, high, half-high and low bushes. Low-bush blueberries are closest to their wild relatives and tend to have smaller berries with an intense flavour. They are rarely used in cultivation, instead high and half-high berry bushes are the norm. A mature high-berry bush grows to 1.8m (6ft) tall and around 1m (3ft 2in) wide, so these need large pots. Half-high-berry bushes grow to around 50cm–1.2m (20in–6ft) tall and are more suited for small spots and containers.

Blueberries like sun or part shade, a sheltered position, no cold winds and well-drained but moisture-retentive soil.

If you're on heavy clay, dig in sharp grit as well as ericaceous compost before planting. They need very little pruning, a good bush should have one third old, one third middle aged and one third young stems. Prune out anything that is dead, diseased, weak or twiggy or any stems that are touching the ground. Remove a quarter of the oldest growth of mature plants to rejuvenate the plant. 'Duke' is a very reliable upright high-bush with good flavour. If you grow it in a pot it will need to be well fed. 'Blue Crop' is another reliable choice that is widely available. 'Bluegold' is compact in growth and suited to pot culture, with a lovely flavour. 'North Country' has the best wild blueberry flavour. It is a half-high bush and grows to 40cm (16in) tall and 1m (3ft 2in) spread and is best grown in the ground.

Pepper Trees

Zanthoxylum species

POSITION
Full sun to light shade, south-, west- or east-facing

SOIL CONDITIONS
Any good garden soil, fertile, well drained

FLOWERING PERIOD
Early to mid-summer

HARVESTING PERIOD
Early to mid-autumn

HEIGHT
2.5–4m (8–13ft)

POLYCULTURE POSITION
Lower storey

It is not easy to grow spices in the British damp, mild climate where baking hot summers are usually needed but one spice thrives and offers a flavour very hard to get in the shop. The seed-coats of pepper trees, known as Sichuan pepper or Sansho in Japan, have a wonderful, slightly mouth-tingling taste with strong citrus notes and warm, woody overtones. They can be dried or roasted with salt and ground into a condiment, or used in pickles. The young leaves are delicious, like a lemony curry leaf, and can be used in just the same way you would Indian cur-

ry leaf. They are very good added to pickle, but when mature they have a small spine to the underside of the leaf, so they need to be picked young and dried or frozen. En masse the seed produces considerable heat, often referred to as *Ma* in Chinese cooking and works very well with ginger and star anise.

There are over 250 species of *Zanthoxylum*, the main ones grown in the UK are *Z. piperitum* from Japan, the toothbark tree *Z. americanum* (very mouth numbing and used as the name

suggests) and *Z. alatum* from Nepal. They are all large, very prickly bushes that grow 2–5m (6½–16½ft) tall and wide, so they need to be sited with that in mind – their spines make an excellent deterrent if you don't want anyone climbing over your fence. They will grow in most soils and need some sun, though will tolerate partial shade. You can prune them much like a gooseberry bush to create an open middle, best done in late winter.

Perennial herbs are a mainstay of any good edible garden because they work so hard in the garden and in the kitchen. Fresh herbs have a such a different profile to that of the dulled, dry sort you buy in the shops, their flavours ripple and ring through dishes. There is a certain joy, even on a dark winter's night in the rain, in tiptoeing into the garden to gather a handful of this or that to make a dish work. The more herbs you have to hand the more you use them. When a recipe calls for a handful of the mass-produced supermarket sort you know that just a sprig of your home-raised ones will do the same job. Even the little used ones, the bay leaves and English mace, are more valuable than they may seem. Herbs are very good pollinator plants, their nectar and pollen count consistently high on the insect menu, many are evergreen or have good winter structures and their laid-back approach to life – you just have to get an herb in the right spot and they do the rest – means that they make up a low-maintenance backbone to the garden.

The Basics
Upright Herbs

Rosemary

Salvia rosmarinus

POSITION
Full sun

SOIL CONDITIONS
Any good, well-drained soil, can tolerate coastal conditions

FLOWERING PERIOD
Late spring to early summer

HARVESTING PERIOD
Year round

HEIGHT
30cm–1.5m (12in–5ft)

POLYCULTURE POSITION
Glade

Rosemary is a tough thing, droughtresistant, woody when old, beautiful in flower, divine when baked in the sun so its essential oils are released. One way to make a rosemary unhappy is to imagine that a shrub that can easily grow to 1.5m (5ft) tall wants to sit in a 30cm (12in) pot for the rest of its life. If you want healthy, plump leaves you have to recognize its root growth. A mature bush is too much for most gardens; the trick is regular harvesting of young growth to keep the bush compact as it doesn't always regenerate well if you cut back into the woody stuff.

I am very fond of the prostrate rosemary, *Salvia rosmarinus prostratus,* that can be planted on the edge of a wall to tumble down elegantly without taking over. *Rosmarinus officinalis* 'Haija' is a tiny cultivar that can be used in hanging baskets and containers, and is good for balconies and rooftop gardens. 'Primley Blue' makes a very neat bush, up to 60cm (2ft) high, so ideal for a low hedge. Rosemary likes full sun and well-drained soil, though it's not particularly fussy about the type of soil. The bright green rosemary beetle is its biggest enemy – pick them off and squish them.

Sage

Salvia officinalis

POSITION *Deep shade, part shade and light sun* **SOIL CONDITIONS** *Good garden soil, fertile, well drained* **FLOWERING PERIOD** *Early to late spring* **HARVESTING PERIOD** *Year round in mild climates* **HEIGHT** *10–30cm (4–12in) when in flower* **POLYCULTURE POSITION** *Glade*

Try purple sage, tricolor sage, golden sage, variegated sage, or elephant-leaved sage with very long large leaves, perfect for dipping in batter and frying as a snack. I think 'Berggarten' is the best for the kitchen – it has particularly broad, silver-grey leaves and rarely flowers.

Sages like full sun and well-drained soil and a plant doesn't much like hanging out in a pot for too long, so repot regularly, refreshing the soil and root pruning if it's staying in a container. Sage is best pruned in late spring before flowering and never cut back into the woody growth as it won't regenerate.

Winter Savory

Satureja montana

POSITION *Sun to part shade* **SOIL CONDITIONS** *Good garden soil, well drained* **FLOWERING PERIOD** *Mid-summer to mid-autumn* **HARVESTING PERIOD** *Leaves harvested throughout the growing season* **HEIGHT** *30cm (12in)* **POLYCULTURE POSITION** *Glade*

Winter savory is tough compared to its summer sibling and made for the sort of stews, soups and hearty bean dishes that you eat in darker months. But just because it's called winter savory, that doesn't actually mean it's around all winter; by December it has dropped its leaves and hunkered down for the cold. So you need to harvest the leaves by the end of summer to store for winter use. Its flavour is strong and peppery, like peppery rosemary with a hint of thyme, far stronger than summer savory, so you don't need huge amounts.

It grows best in slightly alkaline soil, in very good drainage in full sun and is made deeply unhappy by anything overcrowding or flopping on it. Growing to about 30cm (12in) high, it's good for a path edge.

Upright Herbs

English Mace or Sweet Nancy

Achillea ageratum

POSITION *Full sun* **SOIL CONDITIONS** *Well-drained, slightly poor soil, tolerates coastal conditions*
FLOWERING PERIOD *Mid-summer to early autumn* **HARVESTING PERIOD** *Leaves throughout the growing season*
HEIGHT *45cm (18in)* **POLYCULTURE POSITION** *Glade*

English mace with its strongly flavoured ferny green leaves is not a plant that gets a great deal of press. The leaves can be used fresh or dried and have a sharp, aromatic, slightly astringent flavour that needs to be paired with a fat of some sort. It comes from Italy and Spain and like other yarrows it likes a sunny spot and needs a well-drained position. It will rot off in a wet winter in my experience but if you get the sun and drainage right you have a wonderful plant to play with. In summer it has typical creamy white yarrow flowers on very stout stems. It grows to 30–45cm (12–18in) high and can take a lot of wind, so might be one for rooftops, and flowers persist right the way into autumn, making it a valuable plant for bees. It dries well for winter use and the flowers last a long time in the vase.

Mint

Mentha species

POSITION *Sun to light shade* **SOIL CONDITIONS** *Any garden soil, tolerates heavy clay*
FLOWERING PERIOD *Mid-summer* **HARVESTING PERIOD** *Throughout the growing season*
HEIGHT *30–90cm (12in–3ft)* **POLYCULTURE POSITION** *Lower storey and glade*

Mint plays an interesting role in a polyculture. There's an idea that stands of mints help improve the health of the overall system by deterring pests through its scent. I grow all my mint in pots. I see it this way, either I have to spend hours ripping it out of the ground or have to repot every spring. I love mint in tea, especially 'Black Peppermint' and 'Tashkent' mixed together. For salads I prefer apple mint – the leaves are a little hairy compared to peppermints, but the flavour is mild and delicious. Even the tiniest bit of peppermint tends to overwhelm a salad.

Mints prefer some sun and damp feet, but happily tolerate dappled shade. Mint reproduces through runners so it will quickly spiral round the edge of a pot in one season. The best way to tackle this is to repot in late spring, but frankly you can repot any time and a mint won't mind. If you can, keep potting up, but there comes a time when that doesn't work and once you've got a mature plant, take it out of the pot, discard half of it (or give it away) and replant. Top-dress with mulch to keep in the moisture as dry conditions create small, tough leaves.

Most mints will grow 30–90cm (12in–3ft) high. Apple mint grows 60–90cm (2–3ft) and makes a thick, spreading clump. It certainly could be used as ground cover and it will tolerate deep shade as long as its feet don't dry out. There are a number of common varieties based on how the leaves smell rather than flavour, such as grapefruit mint and pineapple mint which was used traditionally to scent sugar.

Korean Liquorice Mint

Agastache rugosa

POSITION
Sun to very light shade
SOIL CONDITIONS
Any good garden soil
FLOWERING PERIOD
Mid-summer to early autumn
HARVESTING PERIOD
Throughout the growing season
HEIGHT
60–90cm (2–3ft)
POLYCULTURE POSITION
Glade

KLM has fast become one of my favourite plants. It's a tough, bushy perennial with lovely heart-shaped, toothed, pale green, slightly hairy leaves that taste as the name suggests, sweet, liquorice with just a hint of mint. By late summer stems are crowned in tall spires of light purple-blue flowers that last and last and last. These flowers are so loved by the bees that the whole plant often sings. The leaves make an excellent tea, and medicinally it's good for upset stomach and is said to help relieve a headache, which makes it

rather useful for a hangover. The leaves can also be used in cooking. In Korea it's used in fish stews and in Korean pancakes, added late to the dishes to preserve its flavour.

Korean liquorice mint is very easy to grow from seed, preferably started off in a heated propagator, and will flower in its first year. It may self-seed if it finds its happy spot in full sun and well-drained moisture-retentive soils; in part shade the flavour is lost a little. The seed heads are robust and it can be left standing over the winter.

Myrtle

Myrtus communis

POSITION *Sun* **SOIL CONDITIONS** *Any good well-drained garden soil*
FLOWERING PERIOD *Mid- to late summer* **HARVESTING PERIOD** *Leaves through the year,*
berries in mid-autumn **HEIGHT** *1.5–2.5m (5–8ft)* **POLYCULTURE POSITION** *Glade*

Myrtle is a like a miniature bay with glossy, deep green, small leaves no more than 2cm (¾in) long and tiny, white, fragrant flowers that are followed by dark myrtle berries. If you get enough of these you can make a killer liqueur, mirto, but you may be waiting some time in the British climate.

This is a classic Mediterranean plant that likes to bake on a rocky, hot hillside in the bay of Naples or Scilly, but doesn't so much love the damp summers of the Midlands. I get around this by growing it in a pot, where it is sheltered from the worst of the winter weather, particularly the wet, then in summer I move it somewhere it will bake and the leaves will give off their delicious resinous scent. In winter it sits opposite my kitchen sink window and I'm always very grateful for all that shiny green when everything else is sparse.

It needs a large pot eventually because it's a shrub that can grow up to 2.5m (8ft) tall. It can be clipped into a pleasing shape, but that means sacrificing the flowers. The leaves can be used in cooking, particularly good with roasted meats or in soups and stews; use it like bay but expect slightly spicier notes. I use up to five small leaves where I would use a single bay. And if I had to choose between the two, I'd go myrtle for those flowers.

Bay

Laurus nobilis

POSITION *Sun to part shade* **SOIL CONDITIONS** *Good well-drained garden soil*
FLOWERING PERIOD *Mid- to late spring* **HARVESTING PERIOD** *Year round* **HEIGHT** *Up to 7½m (24½ft)*
POLYCULTURE POSITION *Upper storey as a tree, lower storey as a bush*

I have both myrtle and bay in pots. This is the only place I suggest putting a bay otherwise you get a very big tree. A wonderful thing about the noble bay is that is doesn't mind how small its shoes are; you can grow it in any pot bigger than 30cm (12in). Keep it clipped in a pyramid or a lollipop, water it when you remember but it's hugely drought-tolerant. Give it sun or give it shade, it won't mind, but a bit of both is kind and top it up with new compost every spring. Prune in spring and summer, and root prune in early spring if it's outgrowing its pot and you don't want it to get any bigger.

The trees are hardy to -5°C (23°F), but when grown in a pot they are more susceptible to damage from a hard frost which may cause the bark to peel and leaves to crisp. The best action is to do nothing, it will inevitably recover, but if temperature drops below -5°C (23°F) regularly, cover the pot with fleece and move it somewhere protected.

Climbers, by their very nature, are tenacious sorts that will go for it once their roots are happy. There are many adaptations from scrambling to clinging and twining, to suckering and those that pin themselves to others to get to the light that they need. They are an invaluable resource in a small garden because they make the most of space. Climbers are incredibly important habitats for insects; mature specimens offer nesting and resting spots for birds, and when in flower they offer plenty for pollinators too.

Nearly all the good edible climbers need some sort of support to climb up whether that is trellis or wires. It's helpful to use vine eyes or to put the trellis on a frame so that it sits just away from the wall or fence to increase air circulation around the plant and help maintain the wall or fence. Some climbers can be encouraged to grow up other plants, particularly trees. However, competition around the base of a tree for root space means that some effort has to be put in to establish the climber, and you also need to either accept that a rumbunctious climber might take over and smother a small tree eventually or try to mitigate that.

The Basics
Climbers

Brambles

Rubus species

POSITION *Sun to part shade* **SOIL CONDITIONS** *Not fussy*
FLOWERING PERIOD *Mid-summer to early autumn* **HARVESTING PERIOD** *Throughout the growing season*
HEIGHT *Up to 3m (10ft)* **POLYCULTURE POSITION** *Glade*

The common bramble is one of a large family with members, including the raspberry, that spread across the globe. There are tender New Zealand types that come with jewel-toned fruits in orange, red and pink, there are shade-dwelling Chinese species and then there's the British bramble, which is loosely grouped as a microspecies made up of many variations of leaf shape, flower shape and fruit shape. In most species, the stems (canes) are biennial, bearing leaves in the first year and shoot- bearing leaves and flowers from the axils of the first-year leaves in the second. The long and short of this is, prune out year one's growth and you won't get any fruit.

Rubus ellipticus, yellow Himalayan bramble, is a scrambling evergreen that grows to 4.5m (15ft) with fairly upright stems with very attractive pale green, grey beneath, rounded leaves. The stems are covered in dense red or brown hairs and stout prickles which it uses to hook itself up whatever it can. It may need to be tied in to keep it in place. The flowers are small and white and the fruit is yellow with a good raspberry flavour. It fruits best in sun, but will tolerate some shade.

Rubus fruticosus, the bramble or blackberry, has many members, some are definitely worth seeking out. 'Oregon Thornless' has stems to 2.5m (8ft) with white flowers and dark black, juicy fruit. It will need to be fan trained to a south-, west- or east-facing wall. All thornless types need to be tied in. 'Loch Ness' is a common variety with thornless stems, happy enough tied on an east-facing wall with a little shade, fruiting prolifically from late summer to last frost.

'Cacanska Bestrna' is not easy to find, but this Serbian-bred bramble has delightful pink flowers followed by juicy, highly aromatic dark black fruit. Canes reach around 2m (6½ft) and will need tying. Give it full sun on south- or west-facing walls for the delicious berries. 'Silvan' is super prolific with large, delicious berries in mid- to late summer but it's a vigorous grower up, to 2.5m (8ft), with very thorny stems, best on a trellis or pergola.

Rubus henryii is a scrambling evergreen shrub that grows to 6m (19½ft) with very attractive tri-lobed leaves, hairy dark green above, downy grey beneath. The flowers are pink and prolific, followed by small black fruit. This is a good one for covering an unsightly wall, doing best in a sheltered shady spot where it won't dry out. It's not for a huge harvest, but a nice nibble.

Rubus henryii var. *bambusarum* is the bamboo leaf bramble, with very elegant, elongated bamboo-like leaves and with small hooked prickles for climbing. Flower stems are red, flowers are pink and small fruit are dark, shiny black. Again, one to nibble rather than a big harvest but it adds a very elegant tropical air to any garden.

Rubus phoenicolasius, Japanese wineberry or Japanese climbing bramble, is one of the finest of the brambles. Covered in fine red bristles, its curving stems arch elegantly and bear clusters of white flowers followed by beautiful orange-red fruit that taste both sharp and sweet, and delicious. The berries appear in late summer to early autumn, after which the foliage turns a delightful autumnal yellow. It does best in a sunny position in any well-drained soil and can very successfully be fan trained against a wall. Older canes should be pruned out after fruiting and new canes trained in.

Grapes

Vitis vinifera and hybrids

POSITION
Sun to part shade

SOIL CONDITIONS
Good garden soil, fertile, well drained

FLOWERING PERIOD
Mid-summer

HARVESTING PERIOD
Grapes from late summer to early autumn, leaves harvested at any time during the growing conditions

HEIGHT
Up to 6m (19½ft)

POLYCULTURE POSITION
Upper storey, walls, pergolas, arches

Grapes will climb anything and will grow in part shade and on north-facing walls, but the best fruit comes from full sun. Growing for grape production requires quite a bit of work, not just careful pruning but for really good dessert grapes also thinning of fruit and some sort of greenhouse protection. This is all a bit more work that I care for. I have a hybrid grape called 'Brandt'; it's a black grape used for wine, known mostly for its lovely autumn foliage. It produces a lot of small, black, quite sharp fruit that probably make delicious wine, but I stop at jelly and juice. I grow my vine mainly for the edible leaves that I use for stuffing, and also in pickling. Grape leaves help to keep ferments crisp.

Grape vines need something to twine around, usually a network of horizontal wires though they can be trained to grow vertically too. You may need to prune to contain them. If growing for leaf production this can be done as part of the harvest, removing whole shoots that are getting too long and removing the leaves. If you are growing for fruit you need to remove leaves around the grapes in order to encourage them

to ripen. For leaf production remove any flower buds as they appear to maintain vigour.

Pruning is done in early winter and should be over by mid-winter. There are many different methods for training and pruning. I maintain a sort of hybrid of the 'curtain' method where I have two main arms growing off a single stem and cut back all the side shoots to 2–3 buds on each. Each summer I train the new laterals up and around my kitchen window.

Leaves should be picked just before using as they wilt quickly. They can be stored by rolling them in brine, drying or freezing. Stack enough leaves for a meal in a freezer bag, exclude as much air as possible and freeze or dry. If you are drying, spread the leaves out on baking trays or something similar and leave out of direct light. Pack leaves together when they feel crispy and crack when bent. Store in an airtight container. Traditionally leaves are sown together in bunches and dried on a washing line or hanging up out of direct light. Rehydrate by dipping in boiling water for two to three minutes.

Mashua

Tropaeolum tuberosum

POSITION
Sun to light shade

SOIL CONDITIONS
Any good garden soil, fertile, well drained

FLOWERING PERIOD
Late summer to early autumn

HARVESTING PERIOD
*Leaves and flowers in the growing season, tubers
are dug up after first frost and any time during the winter*

HEIGHT
2–4m (6½–13ft)

POLYCULTURE POSITION
*Lower storey, will scramble up other shrubs,
walls, fences*

It took me a long time to come round to the roots of mashua. The leaves and flowers were no problem – they are delightful to look at in growth, particularly with sun behind them silhouetting their loopy lobes and their trumpeting orange and red flowers, and they taste good in salads and salsa verdes, hot and spicy. But those rounded, potato-sized tubers were another thing. Unlike potatoes they are not starchy, but importantly they are high in protein and vitamin C, which makes them a valuable, low-fat, low-carb food for the winter. It also means that without the starch they tend to collapse into mush. Some claim they are best in stews, or parboiled before baking, but I think it's mostly about slow cooking. They need a slow steady roasting rather than being blasted. The flavour when raw is really quite spicy; when cooked there are notes of anise and vanilla and still some of that pepper. They work well with other hot flavours, in stews rich in tomatoes and peppers or slow roasted and then drizzled with a chimichurri or a green sauce made from the leaves if they are still around.

I grow both my plants in large, galvanized containers, but mashua can happily grow in the ground in a sheltered spot. Mashua is the same family as the nasturtium and is very fast growing and happy in any reasonable soil in sun or part shade. I often use them as quick cover crops to fill in the gaps while other slower-growing vines are getting growing. The tubers only truly start to develop when day length shortens with longer nights, around early autumn onwards. There are some day length-neutral varieties like 'Ken Aslett', but I think it's not really an issue that you can't harvest till winter. They definitely taste best after a little frost. Tubers can survive a ground frost, but if you are going to leave them in their tubs over winter, which I always do as I find they don't dry out this way, then put a layer of straw or mulch

over the crown to help protect them. I tend to harvest as and when I need them, but always leave a few big fat tubers to grow for next year.

If your winters are prone to prolonged ground frost then it might be best to dig all the tubers up in late autumn, air-dry them and store them somewhere cool and dry, eating over winter and keeping some to replant in spring. I find they are easiest to start off in pots and plant out once they start to need something to climb. They can grow up pretty much anything they can cling to, though they seem to slip down very smooth canes. Mine climb over other climbers, but they'll happily scramble over a shrub.

Roses

Rosa species

POSITION

Sun to part shade

SOIL CONDITIONS

Any fertile, well-drained garden soil

FLOWERING PERIOD

Late spring to late summer

HARVESTING PERIOD

Flowers during growing season, hips in autumn

HEIGHT

1–6m (3ft 2in–19½ft)

POLYCULTURE POSITION

Lower storey suitable for trees, fences, pergolas, arches and house fronts

Rosa rugosa, the tomato rose, has by far the largest hips, but as it makes a shrub that grows to easily 2–3m (6½–10ft) wide, it's not suitable for most small gardens. I do however think a rose somewhere in the garden is a delightful thing. As long as the flowers are simple, single sorts they provide plenty of food for the pollinators and you can use the petals and hips for jams and sauces. The hips are rich in vitamin C, but do need processing as the internal hairs around the seed must be removed or may cause stomach upsets.

Young rose leaves are very good in tea, with a flavour not dissimilar to black tea. If you want a strong rose flavour for herbal teas, bulk out the young leaves with unopened flower buds which are truly loaded with flavour. Dry them either in a dehydrator or on baking trays out of direct light. Young rose petals coarsely chopped can be added to salads, used to make a conserve or preserved in honey, both of which are particularly good over yoghurt.

There are many rose varieties to choose from. *Rosa filipes* 'Rambling Rector' is a rambler, but can be tied in and trained to climb. It has large heads of small, creamy white, semi-double flowers with a strong clove scent that are followed by clusters of small decorative hips in autumn. It's a huge beast of a rose, growing to 8m (26ft) height and spread in about ten years, so this is not for the faint-hearted but its clove-scented petals make such delicious tea so do grow it if you have space for it to do its thing.

Rosa 'Generous Gardener' is more polite, a semi-evergreen semi-climber or large shrub growing to about 4m (13ft), with beautiful, nodding, pink, open flowers with a strong, delicious old rose scent. It has good orange hips in autumn and is very disease resistant.

Rosa moschata has creamy white, single flowers in wide branching sprays late in summer and then good hips. It has a very strong, delicious musk fragrance making it great for teas. It grows to 4.5m (15ft) tall and is suitable for fan training on a wall, sending over a large arch or pergola or up into a small tree.

Rosa 'Shropshire Lass', another semi-climber, has large, pink blushed single flowers that fade to white with a good myrrh-like scent in early summer, followed by masses of orange hips. It will reach 4.5m (15ft) tall against a tall wall or fence or growing up a small tree, and it's one of the few roses that will tolerate shady areas.

An umbel is a distinct-shaped flowerhead typical of the carrot family, *Apiaceae*. It is made up of a cluster of many tiny flowers in which the stalks spring from a common centre to form a flat or curved surface. The result is that the stalk looks rather like a parasol or sunshade, hence the name umbel which means umbrella in Latin. Umbel flowerheads can be tightly packed and the size of dinner plates or delicate, airy arrangements. They can be flat topped or almost round. Their joy is that they hold their looks as seeds and beyond, for once the seed has shattered the umbrella ribs are still quite something with a morning hoarfrost on them.

The carrot family is huge and has many members. Here we look at the reliable biennials and perennials that will do three things: give you something to eat or use medicinally, give the pollinators a feast and finally give you a permanent herbaceous layer to plant around. One of the simplest, but most effective tricks of good design is repetition, but in a small garden using the same plant over and over again for good looks is a little boring. The joy of umbels is that each member can be distinct, but they associate together so you get a garden that feels coherent and has a rhythm that pleases the eye.

The Basics
Umbels

Herb Fennel

Foeniculum vulgare

POSITION
Sun to light shade

SOIL CONDITIONS
Any good well-drained garden soil

FLOWERING PERIOD
Late summer

HARVESTING PERIOD
Leafy greens throughout the growing season.
Seeds in early autumn

HEIGHT
1.2–2m (3ft 10in–6½ft)

POLYCULTURE POSITION
Glade

Tall and about 50cm (20in) wide when it matures, this is best at the back of the border. Its feathery foliage in spring is a welcome sign that warmth is on the way. By late summer it is claiming its spot with acid yellow flowers and then thick seed heads in autumn. The purple form, 'Purpurescens', adds dusky bronze tones to the garden. I mostly use it for height and have perhaps more than needed, but I'm a huge fan of fennel seed tea which is very good for digestion, and although the flower stems toughen, the smaller foliage around the base tends to stay soft all summer.

It will self-seed freely into gravel and free draining soils, and tolerates light shade. Mature plants very much resent being moved so either propagate new stock in spring from seed or dig up a young seedling and move that. Fennel will not suppress weeds in any way, its growth is all about height, so it may need a low-growing companion, but not one that starts too early that will out-compete new growth. Wild strawberries are ideal.

Angelica

Angelica species

POSITION
Light sun or semi-shade
SOIL CONDITIONS
Deep moist, fertile soils
FLOWERING PERIOD
Early to mid-summer
HARVESTING PERIOD
Seed in late summer, stems and leaves when young
HEIGHT
1.2–1.5m (3ft 10in–5ft)
POLYCULTURE POSITION
Understorey

These are tall, statuesque handsome umbels that are often biennials, meaning they form a rosette in their first year and flower in their second year, where they will liberally set seed. They are majestic, but tend to take up quite a bit of space and make the soil around them rather dry. There are 18 or so angelicas which have varying degrees of deliciousness. The seeds, stems and roots can all be used medicinally for digestion, as a tonic, for poor circulation and as a diuretic. The root is chiefly used in its first year of growth before it flowers and becomes woody.

Although *Angelica gigas,* the Korean angelica, probably wins on looks with its dusky purple-pink stems and flowerheads, it's *Angelica archangelica,* Norwegian angelica, that you want to grow for eating. *Angelica sinensis* is worth noting for its strong medicinal properties. Known as female ginseng and used widely in Chinese herbal medicine, it has been used for thousands of years to treat female issues, including period pain and menopause and acting as a blood tonic. It's a useful umbel for dappled shade, such as under fruit trees.

Angelica archangelica, grows to 1.5m (5ft) tall and 80cm (2½ft) wide. Plants are biennial, hardy and will readily self-seed. They grow happily in any soil type, even quite acidic, in sun or semi-shade as long as the soil is moist. Stout flower stems are blushed reddish pink with contrasting yellow-green, rounded flowerheads which will be swamped with bees. All parts are edible – the leaves are good cooked, with a distinct anise flavour, and young ones can be used raw, delicious in a mixed salad. The young stems can be peeled and used like celery in spring, and they are most famously candied, turning bright green, and used on cakes and in confectionary. The seeds can be used in teas and used to flavour spirits and liqueurs.

The other delicious angelica is 'Ashitaba' from Japan, known for profound health benefits and reputed to be one of the famed foods of the Samurai soldiers that promoted long heath.

Modern science suggests that it may have powerful anti-cancer properties. Whether it increases longevity, only time will tell, but it does taste wonderful, sweet and pungent, herby with a hint of perfume and none of the liquorice that is typical of many angelicas. All this deliciousness means that it is rather loved by slugs, but once you get it to a mature plant it soldiers on. It is hard to get hold of, you have to get fresh seed, but the internet has sources. Seed should be sown in early spring in heat and moved outside once frost has passed, with young plants grown in pots until large enough to take on a slug. It is not fully frost hardy, so needs shelter, preferably in dappled shade with damp feet. The plant grows to about 1.2m (3ft 10in) tall with large, glossy green leaves, and it's perennial, self-seeding late in the season around mid- to late autumn, although in damper climates the seeds tend to rot off before ripening. The stem is stout and bleeds a violent yellow when damaged or when a leaf is picked, but it's this that holds the regenerative medicinal properties and a new leaf appears within a day of picking. It's particularly good chopped raw into sushi rice, and also pickled.

Mitsub or Japanese Parsley

Cryptotaenia japonica

POSITION *Part shade* **SOIL CONDITIONS** *Any good garden soil. Loves leaf mould*
FLOWERING PERIOD *Late summer* **HARVESTING PERIOD** *Throughout the growing season*
HEIGHT *Up to 1m (3ft 2in)* **POLYCULTURE POSITION** *Understorey*

A parsley substitute from Japan, this likes the shade and is reliably perennial once it gets its feet down. It has long, tender leaf stalks, up to 40cm (16in) tall, that the slugs are rather fond of. Each stalk bears three heart-shaped leaves, so this looks like a slightly sparse flat-leaved parsley. The fragrance and flavour of the leaves are somewhere between chervil and parsley and it's widely used as a garnish in Japan. The roots are also edible as well as the seeds that are used as seasoning to impart a celery-like flavour. A purple-leaved form is more slug resistant and very elegant when mixed with my dark pink astrantias. It grows to 60cm (2ft) wide.

Wild Korean Celery

Dystaenia takesimana

POSITION *Part shade to full sun* **SOIL CONDITIONS** *Prefers good garden soil but copes with poorer conditions* **FLOWERING PERIOD** *Mid-summer* **HARVESTING PERIOD** *Early spring to late autumn, year round in a sheltered spot* **HEIGHT** *Up to 1.5m (5ft)* **POLYCULTURE POSITION** *Understorey and glade*

Tough as old boots and very hardy, this is another useful perennial umbel. I've read that it was once fed to pigs – lucky things as it's tasty. You can use leaves and stalks in soups, stews and anywhere you might use celery. It has a slightly more perfumed flavour than our leaf celery, but is also a little sweeter. From mid-summer to early autumn it has large, open umbels of white flowers. The leaves are deeply divided and it will happily grow up and through other perennials. Most importantly, it seems little bothered by slugs.

Caraway

Carum carvi

POSITION *Sun to light shade* **SOIL CONDITIONS** *Any good garden soil, likes moist conditions* **FLOWERING PERIOD** *Early to mid-summer* **HARVESTING PERIOD** *Seeds, mid- to late summer. Leaves any time during the growing season. Root in its first autumn* **HEIGHT** *40cm (16in)* **POLYCULTURE POSITION** *Glade*

Caraway is much loved by bakers of bread for its strong, clean-flavoured seed and works particularly well with rye flours. Homegrown stuff certainly packs a punch and just three or so plants will provide plenty of seed for the home baker. A classic biennial, in the first year it looks like a small carrot and neither takes up much space nor demands too much. In the second year airy umbels of white or blush pink flowers appear weaving through the garden and it very much holds its own through summer. It grows to about 40cm (16in) high and is not particularly fussy about soil, but pre-fers slightly damp conditions in full sun to part shade. It has a considerably bigger tap root than you might expect which can be harvested in the first autumn (though that's the end of the plant). If growing in a pot, this needs to be more than 20cm (8in) deep. Some say it does best if cut back in its first autumn, but I've never bothered and always get plenty of flowers and more than enough seed. As it's biennial and doesn't tend to self-seed that well in my experience, save some seed for new plants and sow while the seed is fresh in autumn as a cold winter chill seems to aid germination.

Sweet Cicely

Myrrhis odorata

POSITION *Sun to light dappled shade* **SOIL CONDITIONS** *Any good moist, well-drained garden soil* **FLOWERING PERIOD** *Late spring to early summer* **HARVESTING PERIOD** *Leafy greens throughout the growing period. Seeds in mid-summer* **HEIGHT** *Up to 1.2m (3ft 10in)* **POLYCULTURE POSITION** *Understorey*

This deserves a spot in every garden. It has slightly hairy, much divided, delicate, fern-like mid-green foliage and in late spring stems are topped by frothy white umbels of flowers followed by pointy seed heads that eventually turn jet black. If cut back hard after flowering it will do the whole thing all over again at some point in the summer. Every part of this plant is edible and tastes sweetly of aniseed, so is often used paired with acidic fruits such as rhubarb to naturally sweeten them. The leaves can be eaten raw or cooked. There's some breeding underway to create a cultivar with slightly less hairy leaves so that it's more suitable to eat raw as it does feel a bit like chewing on a paper tissue if you eat too much. However, the seeds and roots are delectable. Young, fresh green seeds, before they go yellow to black, taste just like sweet aniseed balls. The roots can be peeled and roasted, mellowing the aniseed to a parsnip-like sweetness.

Sweet Cicely will self-seed readily. Rather than letting it take over the damper, shadier part of my garden where it is happiest, I dig up the largest plants and harvest the roots which are best mid-spring. Bees love this plant, and both the flowers and foliage work well in the vase.

Coriander

Coriandrum sativum

POSITION *Sun to part shade* **SOIL TYPE** *Any good well-drained garden soil*
FLOWERING PERIOD *Early to mid-summer* **HARVESTING PERIOD** *Throughout the growing season*
HEIGHT *Up to 50cm (20in)* **POLYCULTURE POSITION** *Understorey and glade*

Coriander is a lovely annual in flower, slender and waving with a sparse umbel of white flowers followed by bright green seed that eventually fades to the familiar buff. All parts are edible from the leaves to seeds to roots which can be dried and used as condiments or used fresh in curries, pickles and ferments.

To get it to self-seed successfully you need a long succession of flowering with the seed ripening between late summer and early autumn. This is contrary to how most garden books will tell you to grow coriander as people are generally trying to maximize leaf production and in hot months the plants will bolt to flower quickly. So the usual advice is not to sow into the heat, but you want that seed so start sowing direct in mid-spring to have seed ready by late summer. After many years my polyculture is quite dense so I allow coriander mostly to grow as an understorey crop in large pots on the patio and often manually scatter ripe seed around the pots. If I have a bountiful harvest of seed, I scatter some liberally around the garden to chance my luck, and I nearly always save some seed and store it, in part for the kitchen and partly as a backup if no seedlings come up by themselves.

Chervil

Anthriscus cerefolium

POSITION *Sun to deep shade* **SOIL CONDITIONS** *Any good garden soil*
FLOWERING PERIOD *Late spring to early summer* **HARVESTING PERIOD** *Leaves during the growing season*
HEIGHT *15–30cm (6–12in)* **POLYCULTURE POSITION** *Understorey and glade*

Chervil is such a pretty plant, a delicate filigree of fine green leaves and lacy white flowers. It does best in well-drained but moisture-retentive soils and doesn't like drying out over the summer, but let it self-seed and it usually finds its perfect spot; for me this is usually the path edge where it can grow in the shade of taller plants.

With successional sowings, it is possible to have leaves all year round as chervil is hardy to around -10°C (14°F). Allowing it to self-seed means you usually have plants germinating in late summer and early autumn once the seed has ripened and dropped, growing over winter to flower in early summer. The leaves taste of aniseed and are good in winter salads, and often used as a spring tonic with cleavers and nettles as they have benefits for the liver and kidneys. It is said the leaves are good in a poultice to help aid slow-healing wounds.

These are plants that will cover bare ground quickly, keeping down weeds and making sure the soil remains protected. All of them have a habit of taking over or self-seeding a little too readily, but they are easy enough to uproot. If you can't get to the garden that often, or have a spot that you would like to grow a little wild, establishing these beneath trees or around the base of shrubs is an ideal solution.

The Basics

Ground Cover

Strawberry

Fragaria species

POSITION
Sun to part shade
SOIL CONDITIONS
Good garden soil, fertile, well drained
FLOWERING PERIOD
Late spring to early autumn; intermittently into autumn
HARVESTING PERIOD
Early summer to early autumn
HEIGHT
10–20cm (4–8in)
POLYCULTURE POSITION
Understorey, glade

The garden strawberry likes space, light and high fertility to get maximum production. They are best grown in their own patch, preferably in rows with all the traditional methods. Modern production has moved strawberries from the ground to aerial growing at waist height for ease of picking and the strawberries hang over the edges, and this method could be used for small spaces on a sunny wall or fence or in hanging baskets. You would need to give the plant a bigger pot than you might imagine, 30cm (12in) diameter, and water every day throughout the growing season.

My solution is to have a mixture of a few summer-fruiting strawberries (harvesting from late spring to mid-summer) and many more perpetual strawberries that start fruiting in mid-summer and go on into early autumn. Summer-fruiting strawberries are by far the most prolific. Perpetual tend to produce fruit in bursts throughout the season but they are little more laid-back in their demands. I have learned to allow strawberries to move themselves to new ground; being hungry sorts they are better at understanding which patch is good for them than me. When the mother plant looks tired, but the runners are doing well, I pull up the mother plant and let the cycle go on its own merry way.

I grow wild strawberries too, *Fragaria vesca,* and alpine strawberries, *Fragaria vesca semperflorens.* Both do exactly as they please, and

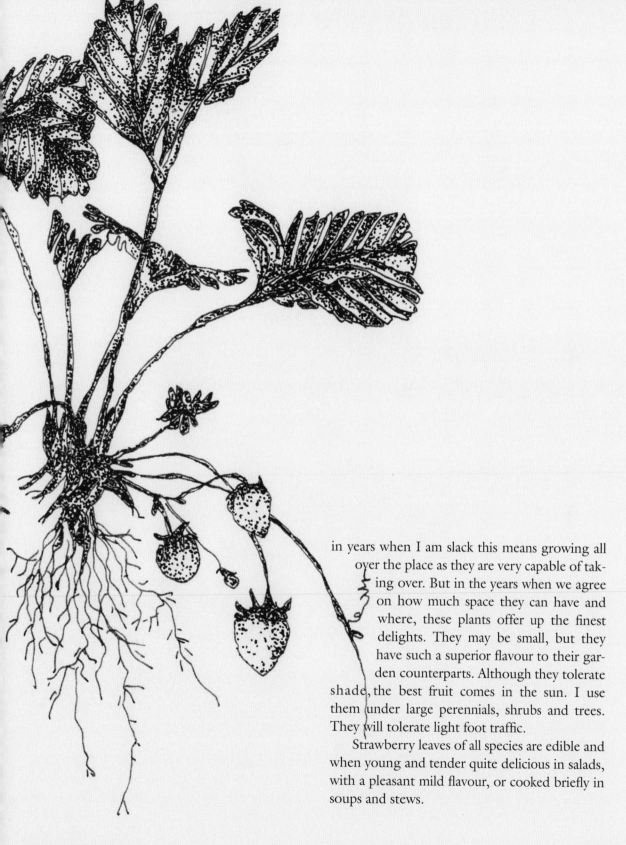

in years when I am slack this means growing all over the place as they are very capable of taking over. But in the years when we agree on how much space they can have and where, these plants offer up the finest delights. They may be small, but they have such a superior flavour to their garden counterparts. Although they tolerate shade, the best fruit comes in the sun. I use them under large perennials, shrubs and trees. They will tolerate light foot traffic.

Strawberry leaves of all species are edible and when young and tender quite delicious in salads, with a pleasant mild flavour, or cooked briefly in soups and stews.

Chinese Artichoke or Crosnes

Stachys affinis

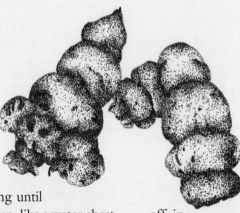

POSITION
Sun to part shade
SOIL CONDITIONS
Good garden soil, fertile, well drained
FLOWERING PERIOD
Mid- to late summer
HARVESTING PERIOD
After first frost in autumn and then throughout winter
HEIGHT
30–45cm (12–18in)
POLYCULTURE POSITION
Understorey

Crosnes are tiny tubers that look just like fat maggots, around 2cm (¾in) long, which sounds revolting until you taste them. They are crisp, like a water chestnut with a nutty, earthy, apple-like flavour that is so good in salads. They keep their crispness when cooked too, briefly stir-fried or boiled. And they are better still lacto-fermented as a pickle.

The top parts look just like woolly mint, with the tubers produced in a string at the end of each rhizome at the end of autumn. I have grown them in very large pots successfully, but you do need to watch out for vine weevil that will make mincemeat out of them.

Plants need full sun and every attempt at growing them in shade has failed; they need well-drained soil and will rot off in very wet conditions. They do best if you mulch them thickly with homemade compost in spring; this will make harvesting much easier. If the plants become too crowded the tubers become smaller but harvesting usually solves this problem as you never get all the tubers up. You can harvest at any point from autumn onward. The tubers don't tend to store well out of the ground, so it's best to harvest as you need these – soak the tubers for a bit before scrubbing!

Oregano

Origanum species

POSITION
Sun to part shade
SOIL CONDITIONS
Any soil, not fussy
FLOWERING PERIOD
Mid-summer to early autumn
HARVESTING PERIOD
Year round
HEIGHT
20–30cm (8–12in)
POLYCULTURE POSITION
Understorey

I use a lot of golden oregano, *O. vulgare* 'Aureum', in the garden. To call it ground cover is perhaps stretching it a bit, but for smaller spaces it will spread nicely and keep the weeds down. It must be grown in free-draining soil as it will rot off in wet winter conditions. It will tolerate part shade but does best in full sun. *O. vulgare* has numerous cultivars with variegated, green or golden leaves, it will self-seed readily and can be sown in spring, though if anyone has a plant they'll happily give you a division.

The leaves are not as flavoursome as the slightly taller sweet marjoram, *Oregano marjorana*, which is the one that is commercially used in pizza and tomato sauces, but in my experience that needs a very warm spot in full sun or a pot. Another fine-flavoured one, Greek oregano, *Oregano onites,* is the one for tomato sauces, but this has to be grown in a pot in very free-draining soil and kept out of winter rain.

Dwarf Raspberry

Rubus taiwanicola

POSITION *Full sun to part shade* **SOIL CONDITIONS** *Well-drained soil*
FLOWERING PERIOD *Mid- to late summer* **HARVESTING PERIOD** *Year round*
HEIGHT *10cm (4in)* **POLYCULTURE POSITION** *Understorey*

This scrambling, tiny dwarf raspberry covers some ground in good conditions and makes a dense mat of fern-like foliage with white flowers in summer and strawberry-like, juicy fruit in early autumn. It needs to be grown in gritty soil otherwise it will rot off over winter. It is not easy to find, alpine specialist nurseries are your best bet.

Bellflowers and Harebells

Campanula species

POSITION *Sun to part shade* **SOIL CONDITIONS** *Good, slightly acid, garden soil*
FLOWERING PERIOD *Mid-summer to early autumn* **HARVESTING PERIOD** *Harvest new leaves throughout the growing season* **HEIGHT** *30cm (12in)* **POLYCULTURE POSITION** *Understorey*

Trailing and alpine bellflowers, *Campanula cochlearifolia, C. portenschalgiana,* and *C. poscharskyana* are mat-forming or trailing evergreen species often found growing in walls or cracks of paths and other very free-draining places. All three produce funnel-shaped pink, purple or white flowers that are much loved by bees. They grow to 30cm (12in) tall and wide and will readily self-seed. All parts are edible from the flowers to the sweet roots. The leaves and stems are best picked very young and you won't get a huge harvest, these are more of a forage food, but the flowers are sweet and delicious and well worth adding to salads, and they do make very good ground cover.

If you want to introduce them to cracks in walls or paving, then mix the seed with sand in spring and sprinkle them about liberally. When they finish flowering, you can cut them back hard and they will reflush with new leaves and often a second smattering of flowers.

Other Plants That
Can Be Used As Ground Cover

Sun to Part Shade

<u>Apple mint</u>

<u>Lemon balm</u>

<u>Aquilegia</u>
flowers are edible and quite delicious

<u>Daylilies</u>
all parts are edible, the flowers, unopened buds, young shoots
and even the roots, eat in moderation

<u>Violets</u>
flowers and leaves are delicious in salads

<u>Solomon's seal</u>
not strictly ground cover, but it will cover ground
creating a clump, works perfectly under apple trees in my garden

<u>Hostas</u>
young, unfurling leaves are best

<u>False strawberries</u>
jury is out on this one, because they taste nothing like strawberries
and they are very rampant, but the crunch of their berries
is delicious and they work superbly in a salad, where they absorb
the dressing in a particularly pleasing manner

Shade
<u>Sorrel</u>

<u>Wood sorrel</u>

<u>Lesser stitchwort</u>
young leaves and shoots are perfect
lightly steamed, like spinach, best in early spring

<u>Wild garlic</u>
if you don't harvest hard,
it will take over your garden!

<u>Solomon's seal</u>

<u>Violets</u>

This group represents the wilder, more adventurous side of eating, plus plants that I rely on to add structure, drama and architectural presence to the garden. They are reliable perennials that need little more than a good start to life and they'll look after themselves.

The Basics

Ornamental Edibles and Pollinator Plants

Sedums

Hylotelephium species

POSITION
Sun or shade

SOIL CONDITIONS
Good garden soil, fertile, well drained

FLOWERING PERIOD
Mid- to late summer

HARVESTING PERIOD
Young leaves in spring to summer

HEIGHT
60cm (2ft)

POLYCULTURE POSITION
Understorey

I probably rely on sedum a little too much, but I never tire of those flat plates of star-shaped flowers in late summer that turn into such good rust and burned umber seed heads for the rest of winter. Once in flower they are covered in bees and butterflies and repeating them throughout the garden is often the one note that holds the garden together come late summer. I started off by growing 'Herbstfreude' or 'Autumn Joy' which has such good bright pink flowers, but the plant has no shame about flopping open and bearing its midriff, which is not only ugly but takes out its neighbours, meaning a lot of staking or cutting back to keep it upright. Best to leave that one to larger schemes and choose one of the smaller varieties such as 'Matrona'. She may not have such vibrant pink flowers but does have very stout upright growth which comes in dusky pinks and smoky purples with an almost pigeon grey dusting. It's a truly good look.

Before flowering, sedum leaves are good to eat, with a slightly bitter, cucumber flavour. Deliciously crunchy, a few leaves finely chopped through salad on a hot summer's day is quite something. You won't ever eat masses of them, but they'll bring other joys to the garden and for one won't garden without them.

Cardoon

Cynara cardunculus

POSITION
Sun to light shade, doesn't mind strong winds
SOIL CONDITIONS
Good garden soil, fertile, well drained
FLOWERING PERIOD
Late summer to early autumn
HARVESTING PERIOD
Leaves in summer
HEIGHT
Up to 2m (6½ft)
POLYCULTURE POSITION
Understorey, glade

The cardoon is the wild, lanky cousin of the globe artichoke. It is much taller, 2m (6½ft) or more to its flower spike which ends in a very spiny, tiny globe with a pink, punkish haircut and very spiky silvery leaves. The unopened flowerhead can be eaten just like a globe artichoke, but needs to be picked before the prickles become lethal, when it is the size of a walnut. The other edible bit is the midribs of the leaves, but these need to be blanched before you can eat them otherwise they are so bitter and tannin rich you wouldn't get past more than a mouthful. The easiest way to blanch them is to put a dustbin over the plants when the leaves are first appearing, and blanch them like you would rhubarb. Or you can earth them up as they grow, which would give you stems to eat by mid-summer. Either way they have to be forced to grow in the dark to sweeten them.

If you get the blanching right, you can peel the stems like celery and eat them raw dipped in olive oil, or boil them and dress them with garlic and salt. They should have the texture of cooked celery and the nutty, earthy flavour of a globe artichoke heart. It's a wonderful plant for a wilder bit of the garden, it will grow on pretty much any soil and the bees truly love it, but if I was limited to space, I'd choose the globe artichoke first.

Globe Artichoke

Cynara scolymus

POSITION
Sun

SOIL CONDITIONS
Good garden soil, fertile, well drained, but moisture retentive

FLOWERING PERIOD
Late summer to early autumn

HARVESTING PERIOD
Unopened seed heads mainly in mid-summer

HEIGHT
1.2–1.8m (3ft 10in–6ft)

POLYCULTURE POSITION
Understorey and glade

Globe artichokes are familiar to most both as a culinary delight and a fine architectural silver-leaved plant for the garden. They are of course reliably perennial, but the best heads come from young plants, split every two or three years in spring before flowering to keep production up. Artichokes easily reach 1.5m (5ft) tall and 1m (3ft 2in) wide. They need to be spaced 60–90cm (2–3ft) apart, but aren't bothered by smaller plants being nestled right up close.

There are far more varieties than the ubiquitous 'Green Globe' which is possibly the least hardy. Smaller-headed 'Vert de Laon' is incredibly hardy and has a much better flavour. The best ones for looks are the purples, 'Sicilian Purple', 'Violetta Di Chioggia' and 'Romanesco' which is the latest to fruit but arguably most delicious. Globe artichokes are a doddle from seed and will flower in their first year if started off early indoors under heat. Seed-grown plants tend to be pricklier, and certain Sicilian varieties are lethal, but their flavour unsurpassable. Growing from seed allows you to get a variety of flavours and lengthens the harvests in one garden. Between the allotment and home, I grow ten plants, which is perhaps excessive, but I like preserving the hearts.

Harvest the heads when they are golf-ball sized when the top of the flower stem is still tender enough to peel. You should get a second flush of smaller secondary heads below. They're not fussy about soil, as long as it is well drained, and there's sun. Protection in harsh winters may be necessary. I use the old flowering stems to cover them, packing with straw and covering with low tunnels to keep the worst of the rain off as rot is as much a problem as hard frosts.

Hostas

Hosta species

POSITION

Full shade to part shade to light sun

SOIL CONDITIONS

Fertile, well drained, moisture retentive

FLOWERING PERIOD

Early to mid-autumn

HARVESTING PERIOD

New shoots in early spring

HEIGHT

10–50cm (4–20in)

POLYCULTURE POSITION

Understorey

Hostas divide people – you either love them and are prepared to go into battle with slugs to get thick luscious leaves, or you can't see the point. As food they make sense to me, and if you are gardening somewhere shady and aren't overrun with slugs and snails or prepared to take on that fight, they are very useful.

You can only eat hostas when their growth is young in spring. By the time the leaf has completely unfurled it has become bitter to keep the slugs at bay. The emerging leaf spike can be eaten raw. It's sweet and crunchy with a hint of lettuce and the young, just unfurled leaves can be used just like spinach. It's a very mild, slightly sweet green that is universally liked in my experience, even by kids. In Japan, their native hostas are considered a wild green or *sansai*. I believe they mostly eat *Hosta montana* and *H. longipes*, but evidence suggest that all hostas are edible. The following are known to be good: *H. plantaginae, H. sieboldii, H. sieboldiana, H. undulata* and *H. ventricosa*.

However, be aware that hosta growers are not generally cultivating their plants as vegetables and may use pesticides, fertilizers and other chemicals that are not suitable for eating. Either buy your plants from organic growers or wait several seasons to eat off commercially grown plants.

Hostas can be grown in part and full shade as well as full sun. In full sun they need moist soil; you often see them growing successful by ponds and streams. The worst slug damage is always when the leaves are just emerging. One hosta grower once told me that they start their slug pest programme on Valentine's Day to ensure that no emerging spikes were damaged. A mature clump of hostas can reflush from an initial cutting, but with younger plants a good method is to cut one half of the clump one year and the other half the next. Hostas in pots need regular feeding, those grown in the ground just need mulching with well-rotted homemade compost in autumn.

Udo or Japanese Asparagus

Aralia cordata

POSITION

Deep to part shade

SOIL CONDITIONS

Good garden soil

FLOWERING PERIOD

Mid- to late summer

HARVESTING PERIOD

New shoots in spring

HEIGHT

Up to 1.2m (3ft 10in)

POLYCULTURE POSITION

Understorey

This wonderful herbaceous perennial makes its presence very much known in the garden. It's a large, leafy architectural plant that grows to at least 1m (3ft 2in) tall and wide, dying back over winter. Large panicles of white flowers that attract a wide variety of beneficial insects are followed by small, black, inedible fruit. The bits you want are the young shoots and leafy stalks, which can be harvested up to about 60cm (2ft) in spring. It's a much-prized vegetable in China and Japan where stems are blanched or earthed up to make them more tender.

A giant leafy green doesn't seem like the obvious choice for a small space, but this one earns its keep in plenty of delicious spring greens, and thrives in deep shade. Udo is a woodland plant and as long as its feet are moist it will be happy in almost any soil type. The plant itself is hardy; roots can survive past -10°C (14°F) but new shoots are more tender so plant this somewhere it doesn't get the morning sun in winter as rapid change in temperature in frosty conditions will cause damage. Don't be afraid of its large leafiness either, sometimes using these oversized plants in a small space adds drama and intrigue. Slugs can be a problem with young plants or when a new plant is establishing itself so try to keep them away during the first season, after that the plant will look after itself.

Udo can be eaten raw; peel the stems and soak in water for ten minutes to remove any bitterness, then slice the stems into dishes. They taste green and grassy, similar to asparagus with lemon and a hint of fennel. If you cook them do so very briefly otherwise they turn to mush.

Jerusalem Artichokes

Helianthus tuberosus

POSITION
Sun to part shade

SOIL CONDITIONS
Any soil

FLOWERING PERIOD
Mid- to late autumn

HARVESTING PERIOD
Tubers are harvested over the winter

HEIGHT
2–2.5m (6½–8ft)

POLYCULTURE POSITION
Lower storey

Jerusalem artichokes are well known, their knobbly tubers are an excellent source of inulin (important in gut health) and this is a crop that won't ever die. Once you have them, you always have them. Chokes can make an excellent windbreak for allotments and open spaces, and a good foil to neighbours' unsightly sheds. If they are growing too tall (they can easily reach 3m/10ft) in mid-season, around early summer, pinch out the tips and they will branch out. The flowers, much loved by bees, appear in late summer and they work well in the vase. Leave the dried stems standing over winter for insects.

Chokes like sunny positions but they'll grow in shade – you'll just get smaller tubers. Plant them in spring at a depth of 15cm (6in). There are a few varieties worth hunting down such as 'Dwarf Sunray' which consistently flowers and produces thin-skinned tubers. 'Fuseau' has large, smooth tubers which are easily to peel and clean, with a very good flavour and heavy cropping. They make excellent crisps.

Tubers are always best and easier to digest once they've had a frost on them. I sometimes dig them up when I know a frost is coming and leave them out overnight to get the sweetest flavour. This tends to decrease the inulin as it is turned into fructose, which makes them easier to digest, but not quite as healthful. Inulin, along with a healthy gut flora, can also help to increase the body's ability to absorb calcium. Inulin can be hard to digest initially, causing flatulence, but your body does adjust, so start off by eating little and often. Also try lacto-fermenting the tubers for a delicious and healthful pickle.

Myoga or Japanese Ginger

Zingiber mioga

POSITION
Shade to part sun

SOIL CONDITIONS
Any good, moisture-retentive garden soil

FLOWERING PERIOD
Late summer

HARVEST PERIOD
Mid- to late summer when flowerbuds are unopened

HEIGHT
1m (3ft 2in)

POLYCULTURE POSITION
Lower storey

Myoga is a ginger relative and it is grown for its emerging flower buds that can be chopped finely for their delicate, floral ginger flavour that is delicious in miso, salads, on rice, with tofu or can be pickled. Once your plant becomes established you can also blanch the stems, much like you might rhubarb, and the long etiolated shoots, when they are pencil thickness, are delicious, crisp and taste very much like common root ginger.

Myoga is thought to originate from southeast China but is now found from central to southeast China to the mountains north of Vietnam and south Korea as well as Japan. The majority of UK plants tend to originate from Japan. In its natural habitat, it is found growing in mountain valleys as an understorey plant of deciduous and mixed forest in shady slopes.

The joy of myoga, other than its surprising flavour, is that it only wants to grow in deep shade. For years, I had a single plant that limped along until I realized it truly was getting too much sun and once I moved it into the shade it romped away. Hence it makes sense to plant it somewhere it can spread. Its rhizomes form very dense colonies and it is once this happens that you start to be able to harvest decent amounts of the unopened flower buds that emerge in early summer. If you want plenty of these, it is worth mulching with very well-rotted compost in spring. Don't use semi-rotted compost as this will attract slugs that will make light work of younger plants. Though they tend to bounce back, it will be at the cost to next year's flowers. It tends to do better in slightly acidic to neutral conditions in a pH between 5.5 and 7. If you are on chalk or lime soils, using a little ericaceous compost, perhaps a 50:50 with regular compost, will help establish the plant.

The plant has thin, lanceolate leaves and grows to around 1m (3ft 2in) high and does best in moist, but not waterlogged soil. The latter is particularly important in winter and if in doubt, mulching in late autumn to protect the plant over winter makes sense. You can grow it in a pot, though it will need root to spread and pot-grown

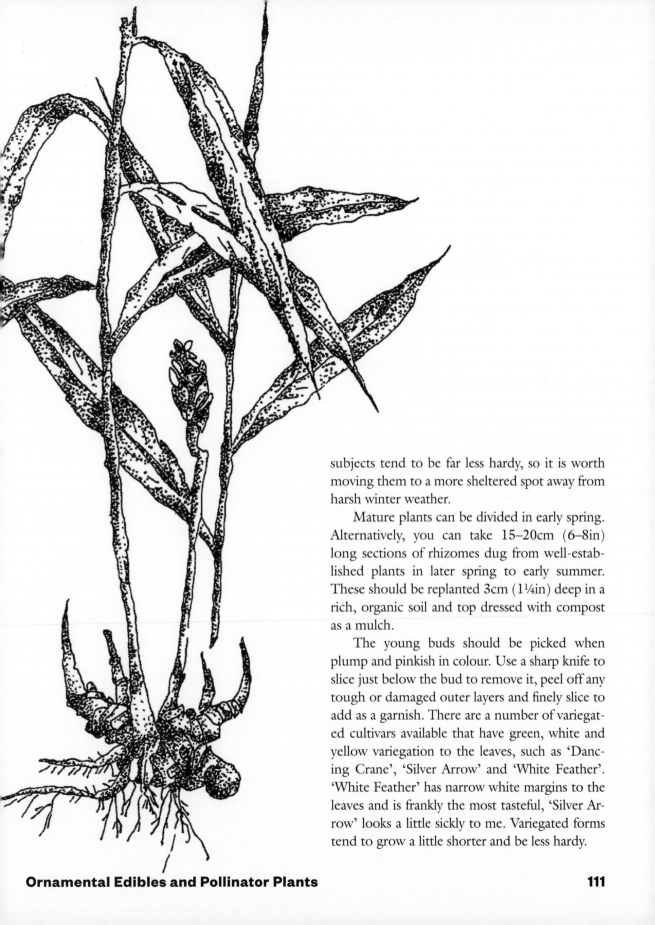

subjects tend to be far less hardy, so it is worth moving them to a more sheltered spot away from harsh winter weather.

Mature plants can be divided in early spring. Alternatively, you can take 15–20cm (6–8in) long sections of rhizomes dug from well-established plants in later spring to early summer. These should be replanted 3cm (1¼in) deep in a rich, organic soil and top dressed with compost as a mulch.

The young buds should be picked when plump and pinkish in colour. Use a sharp knife to slice just below the bud to remove it, peel off any tough or damaged outer layers and finely slice to add as a garnish. There are a number of variegated cultivars available that have green, white and yellow variegation to the leaves, such as 'Dancing Crane', 'Silver Arrow' and 'White Feather'. 'White Feather' has narrow white margins to the leaves and is frankly the most tasteful, 'Silver Arrow' looks a little sickly to me. Variegated forms tend to grow a little shorter and be less hardy.

Oca

Oxalis tuberosa

POSITION *Sun to part shade* **SOIL CONDITIONS** *Fertile, well-drained garden soil*
FLOWERING PERIOD *Mid- to late summer* **HARVESTING PERIOD** *Leaves over the growing season, tubers in autumn* **HEIGHT** *20–30cm (8–12in)* **POLYCULTURE POSITION** *Glade*

This Andean root vegetable is still relatively new on the scene, though it already has many fans. It grows to 30cm (12in) tall and around 40cm (16in) wide with typical oxalis leaves resembling clover. These have a sharp, acidy lemon flavour and can be eaten in moderation but the plant is grown for its tubers which are eaten raw or cooked. If you want to eat them raw dig them up through winter till spring and leave them in the sun for a few days, as this sweetens their sharp, acid lemon flavour to a very palatable crunchy vegetable, making good crudités or added to salads like raw carrot. Or boil them very briefly to eat like small potatoes, slathered in butter. Their lemony flavour is delightful, particularly after frost, their texture is softer and fluffier than potatoes. Varieties vary in colour from white to yellow, orange, pink, dark red and purple.

Leave some in the ground for perennial regrowth, or if you are in a very cold area dig up some tubers to store over winter to replant in spring after exposing them to light to get them to sprout.

Horseradish

Armoracia rusticana

POSITION *Sun to semi-shade* **SOIL CONDITIONS** *Good garden soil, fertile, well drained*
FLOWERING PERIOD *Late spring to early summer* **HARVESTING PERIOD** *Roots can be harvested any time of year, young foliage during the growing season* **HEIGHT** *Up to 1m (3ft 2in)* **POLYCULTURE POSITION** *Understorey*

Woe betide anyone who puts horseradish in the wrong spot because it is almost indestructible and it doesn't matter how many times you imagine you have dug it all out, you haven't. Horseradish will grow pretty much anywhere, in any moist soil in sun or partial shade. A mature plant can easily reach 1m (3ft 2in) tall and 40–50cm (18–20in) wide, so site it with that in mind. It's not so pretty that you want it centre stage.

If you want good, fat, easy to pull up roots, the best way to achieve this is to grow them in an old dustbin sunk in the ground, filled with all the spent compost that accumulates. Unless you are blessed with fine alluvial soils, you'll otherwise get a very forked, very knotty mass of difficult to peel bootlaces.

When young, the leaves are delicious, sweet and only mildly spicy and very good lightly steamed. Slightly older leaves can be used to flavour vinegars. The slightly less vigorous variegated form has lovely leaves for picking. You can blanch young shoots by covering the plant with a rhubarb forcer or bucket to get very tender leaves, and unlike rhubarb you can force horseradish every spring without hampering its growth.

Rhubarb

Rheum species

POSITION *Sun to part shade* **SOIL CONDITIONS** *Any soil, prefers moist conditions*
FLOWERING PERIOD *Early summer* **HARVESTING PERIOD** *Stems are harvested up to mid-summer*
HEIGHT *Up to 1.5m (5ft)* **POLYCULTURE POSITION** *Understorey and glade*

Rhubarb is incredibly long-lived, but as the plant ages it gets less productive. The best way around this is to divide the crown in winter, discarding any woody bits, and to feed with a thick mulch of homemade compost making sure not to bury the crown as that might cause it to rot. Rhubarb are deep-rooting sturdy plants that like rich, fertile conditions. Although they are tough, they will make thin sticks if neglected so try to incorporate organic matter before planting and mulch in autumn and again in late spring. Rhubarb should always be pulled and not cut as this causes the remaining stem to bleed.

Inevitably rhubarb starts flowering at some point in its life. Unless you are into commercial production, it's not essential to remove the flowerhead. You'll maximize stem production but miss the wonderful, giant panicles of white flowers.

'Champagne', if blanched, has delightful light pink stems, 'Holstein Bloodred' has the most brilliant bloody red stem you can imagine, 'Victoria' is the most well known and if you've inherited a plant, it's probably that – early, with lovely pink stems. 'Mammoth Red' is huge, with each leaf growing up to 1.5m (5ft) long. 'Timperley Early' is one of the first to appear and very delicious.

Yorkshire rhubarb is forced in tunnels with candles to heat them to create those lovely neon pink, tender and sweet stems. Home gardeners can use a forcing pot, bucket or dustbin to exclude light, stuffed with straw to make the conditions a bit warmer inside but you cannot blanch the same plant year in and year out, you'll just weaken its growth. If you love blanched stems (and who doesn't?) grow two plants and alternate.

Tulips

Tulipa species

POSITION *Sun to semi-shade* **SOIL CONDITIONS** *Good garden soil, fertile, well drained*
FLOWERING PERIOD *Mid- to late spring* **HARVESTING PERIOD** *Petals when flowering*
HEIGHT *20–60cm (8in–2ft)* **POLYCULTURE POSITION** *Understorey*

The petals of tulips are edible and make for interesting garnish for spring salads. The flavour varies with the colour from bean-like to lettucey to barely tasting of anything. Clearly you should only eat petals that are chemical free and many tulip bulbs are treated with neonicitinoids, but there are more and more organic bulb producers these days, so good for your salads, even better for your bees.

Tulips look marvellous in mixed schemes, working well with lettuce, spring onions and even baby beetroot, but you don't want to be setting out bulbs every autumn for a few petals for a salad or a handful of blooms for the house, so I suggest you choose tulips that reliably flower each year. Most tulips come up blind (as in don't flower) the following year, others that are left to their own devices come up with smaller flowers the following years, but I like this about them: they are closer to their wilder selves and don't overpower a scheme. I've found some pretty reliable, bulking up slowly and returning year after year. 'Spring Green' has white petals with green stripes, tastes good too and looks lovely with stepover apples, 'Groenland' has greenish white petals edged with rose, 'Artist' has garish golden orange and green blooms, but the three work surprising well together. If you want bigger, bolder blooms try pure white 'Purissima' or 'White Triumphator', orange-scented 'Ballerina', pink 'Mistress', red 'Apeldoorn' or the smaller dwarf multi-headed red tulips 'Red Riding Hood' and 'Fusilier' which are both suited to pots, window boxes and rooftops as they won't get knocked over by the wind. For reliably perennial tulips plant bulbs deeper than you'd imagine, as deep as 25cm (10in) as this will mean they won't get disturbed once they've died back and you can plant around them.

I love this layer of the garden, partly because these plants add an air of chance and surprise. These are a mix of annuals, biennials and short-lived perennials that like to move about self-seeding themselves, sometimes with abandon, to wherever pleases them most. The flipside to all this movement is that you will have to decide which seedlings to keep – you can't just let all this lot do exactly as they please as some will inevitably take over or swamp other plants. One solution is to think of the seedlings not as weeds, but as another crop, harvesting them when they are a useful size for the kitchen.

The Fillers

Tree Spinach

Chenopodium giganteum 'Magentaspreen'

POSITION *Sun to part shade* **SOIL CONDITIONS** *Good garden soil, fertile, well drained but moisture retentive* **FLOWERING PERIOD** *Late summer to early autumn* **HARVESTING PERIOD** *New shoots from late spring onward* **HEIGHT** *Up to 2m (6½ft)* **POLYCULTURE POSITION** *Glade*

This large annual reaches over 2m (6½ft) tall, has mid-green leaves covered in a fine white meal (which disappears when cooked) and a brilliant neon pink flush to the new growth. Each plant has thousands of seeds, so you only need to leave a few to ensure the next generation. Leaves are harvested when the plant is young. Pinching out the top tip as you pick makes the plant branch and provide a second harvest before it toughens up and starts to bolt. Then it will produce hundreds of seedlings to harvest whole when you

weed. I only let seedlings grow to full height toward the back of the border. If, when it's setting seed, it's appeared in a part of the garden I don't want it, I pull up the whole plant and shake it like a wand allowing it to scatter a flurry of seeds. You can collect seed in autumn for winter use indoors as microgreens.

It does best in full sun, in rich, fertile, well-drained soil, but it will tolerate light shade and survive in a wide range of conditions. It can be grown very successfully in a large pot.

Orach or Mountain Spinach

Atriplex hortensis

POSITION *Sun to part shade* **SOIL CONDITIONS** *Fertile, well drained but moisture retentive* **FLOWERING PERIOD** *Late summer to early autumn* **HARVESTING PERIOD** *New shoots from late spring onward* **HEIGHT** *Up to 2m (6½ft)* **POLYCULTURE POSITION** *Glade*

Orach is the European cousin to tree spinach and you pick and harvest it exactly the same way. The species has three times more vitamin A than normal spinach, is mid-green and good for eating but pretty boring to look at. *Atriplex hortensis* var. *rubra* is a reddish purple form. 'Magenta Magic Orach' is a bright and spunky dark red when young and picked for cooking, but dusky and sul-

try by the time it grows up. 'Scarlet Emperor' is another in similar vein. 'Bright Flamingo Orach' is distinctly neon pink and makes for wonderful microgreens. 'Golden Orach' is lime green and looks a little sickly when sitting next to its purple siblings, but on its own is lovely. If you grow a selection of oraches they will hybridize and you may lose some of the more distinct colours.

Borage

Borago officinalis

POSITION *Sun to part shade* **SOIL CONDITIONS** *Any, even very poor soil*
FLOWERING PERIOD *Early summer to mid-autumn* **HARVESTING PERIOD** *Young leaves before the flower bud appears* **HEIGHT** *60cm (2ft)* **POLYCULTURE POSITION** *Understorey*

Borage is native to central Europe but has become a garden escapee in Britain. It is hardy so I often have plants over the winter, albeit straggly ones. Most famous for its flowers and oil from the seed, sold commercially as starflower oil, the flowers are traditionally used in summer cups. They are lovely things and adored by bees, but my heart is for the leaves that have a distinctly salty, cucumber flavour I just adore. Young leaves are rich in potassium and calcium but are incredibly hairy and unpleasant raw. Borage loves to self-seed everywhere and I harvest the young seedlings when they are anywhere from 10–20cm (4–8in) tall and the leaves are vibrant green. I massage these with a little salt, leave them for ten minutes, wash them and sauté them briefly until they wilt.

Delicious as they are, however, you cannot make borage leaves a major part of your diet as they contain pyrrolizidine alkaloid, which can cause liver damage and liver cancer. In truth, though, I don't know how you'd do this, as only the younger leaves, before a plant flowers, are worth eating – a brief spell in late spring and again perhaps in late summer and early autumn if they have self-seeded.

Borage will readily self-seed but you can start it off either indoors or direct in the ground in late spring. It is unbothered by pests or diseases, but the mature plant is large and straggly and tends to lean on its neighbours, so be sure to site with this in mind, harvesting other seedlings for supper. I am quite a snob about borage and only let the white flowering form, *Borago officinalis* 'Alba', flower in my garden.

Vegetable Mallows

Malva species

POSITION *Sun to light shade* **SOIL CONDITIONS** *Any soil* **FLOWERING PERIOD** *Mid-summer to early autumn*
HARVESTING PERIOD *Leaves are harvested throughout the growing season*
HEIGHT *50cm–1.5m (20in–5ft)* **POLYCULTURE POSITION** *Understorey and glade*

Mallows have delicious leaves with a mild, almost buttery greens flavour whether eaten raw or cooked. Simply harvest the size of leaves you desire and the plant continues to grow. The flowers and immature seed heads are also edible; the distinctly nutty seed head makes a great nibble.

There are many mallows – I eat three of them regularly. All are pollinated by bees and are good late summer nectar plants. They are generally pest free but can suffer from hollyhock rust, tiny yellow rusty spots that take over like measles. If this happens it's best to dig out any plants and bin them. You can get around rust to some extent by clearing all plants and their plant debris by mid-autumn to eliminate overwintering spores but it's probably better to give them a rest for a couple of years.

Malva sylvestris, common mallow, is a biennial or short-lived perennial that grows to around 50cm (20in) very quickly. It's a thing of waysides and roadsides, of disturbed grounds and margins with very small, pink flowers and huge leaves. It tolerates dappled shade and full sun and I think of the common mallow mostly as a forage food as it's not quite pretty enough – the flowers are just too small – to warrant a main place in the garden. However musk mallow, *Malva moschata,* is a classic cottage garden plant and if you're lucky enough to be free of hollyhock rust it's a lovely thing for any garden. It's a perennial growing to around 80cm (2½ft) tall and 60cm (2ft) wide, flowering in mid- and late summer with heart-shaped lower leaves and more lobed upper ones with pale pink, saucer-shaped flowers.

Chinese mallow, also called vegetable mallow and cluster mallow, *Malva verticillata* is my favourite. It's a very ancient vegetable hailing from China with large leaves that grow to more than 15cm (6in) across and remain tender enough to eat all year round, even the older bottom leaves. A very pretty cultivar, *M. verticillata* 'Crispa', has frilly, folded leaves that look like a handkerchief. It has such tiny flowers that you might blink and miss them, but the bees don't so it will readily self-seed to appear in sunny, well-drained spots. It grows to 1.5m (5ft) tall, so keep that in mind when deciding which one to leave in, and it has a habit of deciding that the path edge is its favourite place. It has come through many winters as a full-grown plant in my garden; it looks like it's been taken out by frost only to spring back. A biennial, once it sets seeds it quickly starts fading and if it's looking unsightly before the seeds are quite ready, up-end the whole plant and hang it upside down somewhere dry for the seeds to continue to ripen.

Salads and Herbs

Salads are such an important part of the garden to me for many different reasons. Firstly, they are one of the simplest ways to have a huge impact on your carbon footprint for eating. Bagged salad is the first thing to rot in your fridge and comes, even when relatively closely grown to home, at a considerable environmental cost. More often than not our salads come from far too far away. These six are so reliable that whatever the weather throws at the garden, however slack I am with maintenance, I know I will always be able to find a handful of leaves of one or other to make enough for a sandwich or salad for lunch. In the height of summer heat, this lot tend to be either in flower or setting seeds, but although they might be a bit thin on the ground in mid- and late summer, they make up for it by being reliable winter leaves. Allowing them to live out their full life cycle means that there is always spring forage for the pollinators and beneficial insects when the overwintering plants start to flower.

Miner's Lettuce

Claytonia perfoliata

POSITION *Sun to deep shade* **SOIL CONDITIONS** *Any, but prefers moist soils*
FLOWERING PERIOD *Late spring to mid-summer* **HARVESTING PERIOD** *Leaves over winter and early spring*
HEIGHT *15–20cm (6–8in)* **POLYCULTURE POSITION** *Understorey*

Miner's lettuce, so called because it has naturalized at entrances to mines, makes a delicious succulent leaf for salads. It can be harvested as whole rosettes as the stems are as good as the leaf, it can be eaten when in flower, or you can pick individual leaves. Its fresh grassy taste works particularly well with soft cheeses. Once you've let it live out its life cycle, you'll find it pops up all over the place. It particularly likes damp, partially shaded spots, often growing at the base of shaded walls. Sow in autumn or early spring direct or in seed trays. It grows to 20cm (8in) at full height in flower, but in leaf is 10–15cm (4–6in) high, a perfect understorey plant for large pots. It is very fast growing so if you sow in late winter you'll be picking by late spring into mid-spring.

Corn Salad

Valerianella locusta

POSITION *Sun to light shade* **SOIL CONDITIONS** *Any, not fussy*
FLOWERING PERIOD *Mid-spring to early summer* **HARVESTING PERIOD** *Leaves mostly over winter and early spring* **HEIGHT** *15cm (6in)* **POLYCULTURE POSITION** *Glade*

Corn salad or mâche is a fast-growing, cool-season annual which means that you sow it in late summer or early autumn and it grows over winter to give you soft, sweet, slightly corn-flavoured leaves. It is very low growing, only 10cm (4in) tall until it flowers when it bolts up to 15cm (6in), setting seed between late spring and early summer. I started off growing corn salad in pots, but it soon decided that it wanted freedom and now pops up all over the garden, mostly along the path edges. Pretty, tiny, bluish white flowers are much loved by lacewings and hoverflies, making it an important plant for the health of an ecosystem, as the lacewing larvae are such good predators of aphids in early spring. It is unfussy about soil type, but does give the best, fattest leaves in moist conditions. It is worth cloching a few plants over winter for particularly tender leaves.

Parsley

Petroselinum crispum

POSITION
Sun to part shade
SOIL CONDITIONS
Any good garden soil
FLOWERING PERIOD
Early to late summer
HARVESTING PERIOD
Leaves throughout the year
HEIGHT
30–45cm (12–18in)
POLYCULTURE POSITION
Understorey and glade

Parsley loves to be picked, the more you harvest the more it will respond. And it doesn't much love life in a pot as it is very deep rooted with long, strap-like, quite elastic roots that will just keep going down to find the minerals and nutrients it wants. The best-tasting parsley is grown in the ground or in a very deep pot. For years, I bought into the snobbishness about flat-leaved being superior to curly or butcher's parsley. No longer. Flat-leaved is superior for salads, but butcher's parsley has a depth and sweetness of flavour that makes it very good for combining with other rich flavours such as chicken livers and lardons, or try it deep fried, where it turns almost seaweed-like.

Parsley is a biennial that is about 30cm (12in) high in leaf, up to twice that in flower. The flowers are lime green and attractive, not showy, but subtle. If you want your plants to successfully self-seed they need to be flowering by early summer so that the seed is ripe by mid-summer and scattering to germinate in late summer. This gives them just long enough to grow to a size that will survive winter frosts, though you may wish to cloche flat-leaved parsley over winter to keep it in good nick for salads. All those curls in curly parsley means that it is far better at surviving frost, as each fold traps air to keep the leaf from freezing.

Parsley germinates best at temperatures around 22–25°C (71–77°F), when it should take about seven days, at lower temperatures it can take 4–6 weeks. It can be sown all summer long and for long production you should make several successional plantings. To get them to self-seed you want plants that have overwintered to flower in their second year as biennials. These overwintered plants may take up space that you want for other vegetables, but remember that they aren't setting seed till mid-summer, so you can carefully dig up plants in early spring before they flower and move them to a more convenient spot. They will still flower well in part shade.

In a dense polyculture, the seed doesn't always germinate because it gets lost in the understorey, so it may make sense to collect the seed rather than allowing chance too great a play. As a side note, the juice of the stems and leaves is an excellent mosquito repellent and will also relieve the pain of insect bites and stings.

Rocket

Eruca vesicaria subsp. *sativa*

POSITION
Sun to light shade

SOIL CONDITIONS
Any, can tolerate drought

FLOWERING PERIOD
Late spring to late summer

HARVESTING PERIOD
Throughout the growing season

HEIGHT
30–50cm (12–20in)

POLYCULTURE POSITION
Understorey and glade

Rocket is such a lovely addition to salads or on top of pizzas and pastas. You can eat it any stage; it is as delicious in flower as in leaf, though you might have to strip the bottom of the flower stem off as it can get a bit tough. It is very easy to grow and, as its name suggests, fast growing to boot. It will grow pretty much anywhere but does best with a little shade in summer as this stops it flowering too early. It is plagued by flea beetle that loves to nibble little holes in the leaves, but it doesn't change the flavour one bit, so don't worry too much. If you don't eat all the flowers the hoverflies will be thankful and it will self-seed merrily. It does have a habit of cross pollinating with other brassicas, particularly mustards and kales, so you may find in time that you have a rocket that is entirely unique to you or you may have something that is too mustardy for its own good, in which case start again with bought seed. The seed has a natural post-harvest dormancy period of a couple of months, where it will sit in the soil just waiting time out – be patient and it will reappear if you've let it set seed. Rocket flowers waving in a late summer's breeze is one of many joys of this sort of gardening. I've never seen this plant not in the right place, it's an effortless, delightful thing.

Landcress

Barbarea verna

POSITION
Sun to deep shade

SOIL CONDITIONS
Any, but prefers moist soils

FLOWERING PERIOD
Late spring to mid-summer

HARVESTING PERIOD
Leaves mostly over winter and early spring

HEIGHT
30cm (12in)

POLYCULTURE POSITION
Understorey

Landcress tastes very similar to watercress and for that matter looks much the same too, but requires no running water. It is a very low-maintenance plant; sow it and let it flower and rest assured it will be somewhere in your garden for the rest of time. It has searing bright yellow flowers in late spring at which point it goes from being a rosette- forming, low-growing plant to around 30cm (12in) high. Bees and hoverflies love it and seed ripens quick-ly; by late summer it is ready to scatter and you can pull up the bleached seed heads. A very pretty variegated form looks particularly attractive when in flower and makes for an unusual salad addition but it is quite hard to get hold of and if you also have the straight species you may find you lose it as it reverts. Initially, sow seed in modules or seed trays to plant out, or directly into a weed-free bed and then allow it to do the rest.

Watercress

Nasturtium officinale

POSITION
Sun

SOIL CONDITIONS
Wet soils, can grow in water

FLOWERING PERIOD
Late spring to mid-autumn

HARVESTING PERIOD
Leaves during the growing season

HEIGHT
25–30cm (10–12in)

POLYCULTURE POSITION
Understorey

Watercress is familiar enough to need no introduction. It is very easy to grow both in a pond or in a pot. As its name suggests, it does need moisture to grow satisfactorily, though it's surprising how well it soldiers on without it. I grow watercress outside but also in seed trays as a microgreen, indoors or in the greenhouse over the winter for a quick spicy green. I sow liberally over peat-free compost, don't cover the seeds, water well and harvest when 5–10cm (2–4in) tall.

In the garden, I grow my watercress in a recycled metal container, though it inevitably self-seeds into other cracks. I often find it coming up around the paving by the outdoor tap and this tells you a lot about what it likes. It wants to be wet, but it doesn't want to be stagnant and ideally it wants free-draining, but constantly moist conditions like the chalk streams it inhabits in the wild. Whatever pot you grow it in, it will do best if there's drainage, and if it is endlessly being topped up with water. My solution is a baby's bath with a rusted hole in the bottom so it drains, but slowly so the watercress always remains damp. Watercress is spreading and with enough space it will form a dense mat of stolons with a deep, fibrous root system. Although it is perennial, it is often easiest to treat it as an annual to maintain its vigour. If you have a stream or natural pond then it is well worth establishing it, but if cows or sheep are grazing nearby and use the stream the watercress can't be eaten raw as it may become contaminated with liver fluke, a parasitic worm.

Any of the brassica pests, the cabbage whites and aphids, can attack it, so keep an eye out for damage. Either pick off the caterpillars or cover to prevent the female laying. If you get an aphid attack it is usually because the plants are too dry. A strong jet of water will dislodge many aphids, but it often makes sense to just cut the plants back (and bin the infected material), mulch with compost and water well and wait for a reflush of growth.

Flowers

This section could be huge – there are so many flowers that please the wildlife and insects and also create a painterly palette for the garden. I've included these three because they are the ones I use most often as herbal medicine, cut flowers and to entice pollinators or to use in the kitchen.

Poppies

Papaver species, *Eschscholzia californica*

POSITION *Sun* **SOIL CONDITIONS** *Any good free-draining soil* **FLOWERING PERIOD** *Early to late summer* **HARVESTING PERIOD** *Dependent on species, but seed is usually ripe in late summer and early autumn* **HEIGHT** *45cm (18in)* **POLYCULTURE POSITION** *Glade*

Papaver rhoeas, P. somniferum and *Eschscholzia californica* are all important medicines that contain various alkaloids including opium. *P. somniferum,* the opium poppy, is the strongest of the three but it is very addictive and should only be used under the supervision of a qualified herbalist. However, the Californian poppy, *Eschscholzia californica,* is safe for general use including for children and makes a very calming, gentle tea that helps aid sleep, restlessness and anxiety, is a diuretic and relieves pain. The whole plant is harvested when in flower and dried for use.

The seed of all three poppies can be used in cooking. It contains none of the alkaloids associated with other parts of the plant and has a distinct nutty flavour. The field poppy, *Papaver rhoeas,* is the only one that you can otherwise eat. The young leaves can be eaten raw or cooked and work just like spinach but they must be picked before the flower bud appears at the centre of the plant. Once the flower bud appears the leaves become toxic and are no longer suitable for eating. This poppy is a common wild forage food of Turkey and the Caucasus.

All three poppies will self-seed freely in the right conditions and do best in fertile, but free-draining soils. Californian poppy in particular likes to sow itself into cracks in walls and paving. There are many named varieties selected for their ornamental value – double flowers, dark purples, light pinks, picotee edges, purple white. The wildlife value is highest in the species and single-flowered forms. Some selections of opium poppies are bred for culinary purpose for bread seed. They don't open their seed heads so readily and are easy to collect, and often sold as 'Hungarian bread seed poppy'.

All poppies dislike root disturbance and if you are raising them in seed trays, pot on into deep 9cm (3½in) or larger containers as quickly as possible and don't let them sit around. Californian poppies make good cut flowers.

Pot Marigolds

Calendula officinalis

POSITION *Sun or part shade* **SOIL CONDITIONS** *Any good garden soil*
FLOWERING PERIOD *Early summer to late autumn* **HARVESTING PERIOD** *When in flower*
HEIGHT *40–50cm (16–20in)* **POLYCULTURE POSITION** *Understorey and glade*

No plant works so hard for so little as calendula, endlessly flowering over a very long season, giving gentle but powerful medicine and feeding the bees whatever the weather. Calendula flowers are known for their antibacterial, antiseptic and vulnerary (wound-healing) properties, and are used to treat complaints such as dry skin, stings, sprains and sore eyes.

Harvest the flowers when they first open, with a little bit of the stalk, preferably before midday when they are their stickiest as the sticky bit is where all the good stuff is. I dry mountains every summer to ward off colds and sore throats and to infuse in olive or almond oil to make a hair conditioner. Huge handfuls in the bath also help heal cracked skin, insect bites or other sores. It's a particularly relaxing way to take your medicine.

Calendula will self-seed readily, flowering better on slightly poorer soil. They do require some sun every day, but are tolerant of most conditions. Sow direct from mid-spring onward. Calendula make a very good cut flower, lasting for a surprisingly long time. The flowers can also be used to predict the weather as they won't open in the morning if it's going to rain later. It's thought this is to prevent the pollen from getting wet.

Nasturtium

Tropaeolum majus

POSITION
Sun or part shade

SOIL CONDITIONS
Unfussy

FLOWERING PERIOD
Mid-summer to early autumn

HARVESTING PERIOD
*Young leaves early in the season, when in flower and seeds late summer
to early autumn before they ripen*

HEIGHT
Up to 1.5m (5ft) scrambling up something

POLYCULTURE POSITION
Understorey climber, walls, fences, through shrubs

Good old nasturtiums, perhaps a little ubiquitous, but reliable, jolly and very useful whether for clothing an ugly fence, in a hanging basket or windowbox or as herbal medicine. The leaves and flowers are wound healing, antibacterial and can be used to make a very hot-flavoured tea that is a good expectorant for coughs and bronchial catarrh, though it does taste very odd.

The flowers are well known for their hot, peppery flavour in salads and work very well chopped up with capers and made into a salsa verde of sorts. Young leaves are equally good and can be used alone or added to salads. The immature seed pods are incredibly spicy and traditionally pickled like capers. Mature seed pods become too tough for this, but you can dry them and grind them for a pepper-like condiment.

Folklore states that nasturtiums grow best in the floor sweepings, which is to say that the poorer the soil the better the flowering as too rich conditions result in all leaf and no flowers. Sow outside direct, in pots or in containers from early spring onward. There are many beautiful cultivars that have been selected for ornamental use. I particularly like the pale-coloured forms such as dusky pink 'Ladybird Rose' or palest cream 'Milkmaid'. I also have a soft spot for 'Alaska' which has variegated leaves that look splattered with white paint. It looks particularly good in a salad and readily self-seeds. If you want to keep your breeding lines pure you will have to sow each year as nasturtiums are a promiscuous lot and will readily interbreed so that your dusky pinks may all be orange the next year!

This is the fun bit – one year you may want a garden rich in salads, another year you might decide to grow only annuals with red or purple leaves. Every year you get to make new patterns, to try new combinations, to enjoy the thrill of the experiment – will it work or not, can you truly get a tomato to be happy growing through a rose bush? (Yes! If you choose the right ones.) The toppings are here to fill in the spaces between your other plants and generosity without overcrowding (so easy to write, so much harder to do) is necessary. These plants are the most labour intensive of the lot; many will need starting off indoors under heat, or raising in modules to plant out. This gives you a bit more control than the fillers that are happily self-seeding themselves about, but it also means they will consume resources: water, compost, pots and lots of your time. It is easy when ordering seeds to get carried away and find your eyes are far bigger that your belly or plot.

Be realistic: a family of four needs two to four courgette plants and even then that will give you more than you know what to do with at the high point of the season, and they will take up a lot of space that could produce armfuls of basil, coriander, summer savory and buttery lettuce heads which might be more useful on a day-to-day basis. If space is an issue I would prioritize flavour over production, ditching three of the courgette plants for a summer squash (firmer than a courgette, but more flavour), a cucumber and a whole lot more salad plants. I wouldn't bother with parsnips but favour more baby beets and I would choose reliable kales and Swiss chard over temperamental bulb fennel, chillies are easier than sweet peppers and tomatoes win over aubergines. Be honest with yourself because there is a great deal of heartbreak in investing hours on a crop that never amounts to much, where there's joy to be found in easier picking.

Finally, it will take time to learn exactly what your household uses in quantities, and time learning to adapt recipes to what's to hand, but a glut is always a waste. No one successfully uses up a glut – hundreds of jars of chutney will remain festering at the back of the cupboard and there's truly only so much lettuce soup anyone can eat. Sow pinches of seed every two weeks of anything that can be grown in succession and choose variety over horde every time. In our 'on demand' culture learning to truly savour something and then let go is a very important lesson. I believe that the garden is best when it gives you a burst of flavour so wonderful you swoon in delight but that offers up some new treat before you could get bored or complacent or just totally tired of peeling and bottling.

The Toppings

Root Vegetables

Root vegetables are earthy things that draw up minerals from deep below to give you the true flavour of your soil. If space is tight eat endless successions of tasty, tender baby versions – parsnips not much bigger than your thumb, beetroot the size of walnuts and use every bit, leaf and all; the same for turnips. If you have more space, grow enough to lay down for stores, for there's nothing as sweet as a beet that has sat in the cool of a clamp all winter. All root crops do best in rich soils where they will grow quickly. If you find that growth is stalling somewhere around the fifth or sixth true leaf, give them a liquid feed with comfrey, seaweed or similar and make sure they have enough water – root vegetables that slow to a stop mid-growth are often tough to eat.

Beetroot

Beta vulgaris

POSITION *Sun to part shade* **SOIL CONDITIONS** *Rich soils* **FLOWERING PERIOD** *Summer onward*
HARVESTING PERIOD *Roots from mid-summer* **HEIGHT** *30cm (12in)* **POLYCULTURE POSITION** *Glade*

Beetroot are a main winter crop. I particularly like them in late winter or early spring when all the sugars from storage have sweetened them to perfection. To have enough to take me through winter, I aim to have around 30–40 or so plants. I grow all sorts and sizes, golf balls for roasting whole and large beets for soups and cakes. I've learned that beetroot really don't do well with early competition so I first sow them quite thickly, then thin in stages, harvesting young beets in early summer to leave very widely spaced beets for maturing for winter storage. At this stage, around early summer into mid-summer, I sow radishes, dill, salads and wild rocket for baby leaf production between the maturing beets. I find this method is the most productive but I recommend you experiment with thinning and spacing regimes to get the size of beets you desire; anything smaller than a golf ball doesn't store well over winter.

You may only want beets for summer production, in which case you can succession sow every two weeks, thinning to around 10cm (4in). You can have a very productive spot by creating a polyculture mix with radishes, lettuce, dill, rocket and baby turnips where you aim to thin everything to 10cm (4in) or so and pick young.

Beetroot are very fussy about good soil in the first stage of their life. If you've found that your beets stall around the three or four true leaf stage, when they are 10cm (4in) or so tall, turning a little red or purple from stress, then it could be because they are low in boron which is key to early good growth. I sow mine into a thin layer of good-quality sieved compost or leaf mould and then when the seedlings are up, I give them an additional feed of seaweed. You can add bo-rax to the ground a week or so before sowing, or fertilize the seedlings as they appear with a very good-quality seaweed feed as boron is often present as a micronutrient. On poor soil it's worth apply fertilizer ten days before sowing; comfrey or organic tomato feed high in potash is ideal.

I love the way beetroot leaves look in a dense polyculture and have come to rely on the famous 'Bull's Blood' with its deep metallic purple-red leaves as a mainstay of autumn colour. People are snobby about its flavour, but I have no beef with it (sorry, I had to go there with that pun) and I love the leaves cooked just like Swiss chard though I'd caution against picking the leaves willy-nilly while the plant is still growing as this weakens its growth and stops beets fattening up.

Beetroot are very easy to start in modules or trays, ideally to plant decent-sized healthy seedlings 10–15cm (4–6in) high into an existing polyculture. Or sow seeds direct – I wait till at least the end of early spring to do this. The rule is that the soil has to be warm enough for you to leave your bare hand on it for a good few minutes. Thin to 7.5–10cm (3–4in) apart for small summer beets, for ping-pong beets thin to about 15cm (6in) apart and for winter storage you can thin anywhere from 20–30cm (8–12in) apart.

'Olympia' is a selected form with uniformly red leaves. 'Egyptian Turnip-rooted' is a very old beetroot that is best for summer use rather than storage; the beetroot is a lovely violet colour and sits on top of the soil like a turnip. They look lovely grown among bright green baby lettuce leaves. 'Sanguina' is a very deep bloody red beet with lovely red midribs to the leaves; it makes an excellent storage beet.

Turnips

Brassica rapa

POSITION
Sun to part shade
SOIL CONDITIONS
Do best in organic rich soils
FLOWERING PERIOD
Second year late summer onward
HARVESTING PERIOD
*Early sowing will give you roots by mid-summer,
but more pickings will be late summer through to autumn*
HEIGHT
30cm (12in)
POLYCULTURE POSITION
Glade

I have no time for large storage turnips, they take up too much space and frankly don't look that fine, but give me a small, perfectly white Japanese turnip and I am a happy gardener. These turnips are so useful. The greens are wonderful young as thinnings or you can eat the tops of harvested turnips, sweated in butter until soft and tender. Young round turnips are lovely, sweet, nutty with a hint of mustard, so delicious glazed in butter or roasted or pickled or even sliced very thinly, marinated in a little vinegar and sugar and served raw. Treat them like radishes: sow thickly, eating the thinnings at every stage. They grow fast and furiously in good soil and light, though they will take a little shade and are a useful intercrop between leeks, onions, pumpkins, courgettes, corn and other slower-growing larger vegetables.

I start sowing turnips around the same time as beetroot, when the soil has warmed up enough to leave my bare hands on the soil and it's not too wet. You can sow successively from late spring to mid-summer and thin in stages to 10cm (4in) apart. The greater the spacing the quicker the roots will develop. If you want to store turnips for winter, sow mid-summer and use maincrop cultivars, thinning to 20cm (8in) apart.

For greens and tops sow from early spring to early summer, you may have to cloche if the weather is cold. Start harvesting when the greens are 10–15cm (4–6in) high and cut 2.5cm (1in) above soil level, then they will resprout. Try eating this green Southern US style as a 'mess o'green'. Melt the greens in butter. Once they start to wilt, add garlic, a little water, a good glug of red wine vinegar, red pepper flakes, black pepper, salt and a little sugar and cook down until soft and fragrant.

My favourite varieties are 'Japanese Snowball' and 'Tokyo Cross' (F1), both lovely pure white, crisp roots with excellent-flavoured tops. 'Oasis' is a melon-flavoured turnip, super sweet and delicious raw. 'Namenia' is a variety bred for fast, tender green tops perfect for harvesting for early spring or autumn greens.

Radishes

Raphanus sativus

POSITION

Sun to part shade

SOIL CONDITIONS

Do best in organic rich soils

FLOWERING PERIOD

Second year late summer onward

HARVESTING PERIOD

Late spring to early autumn, seed pods throughout summer

HEIGHT

15–30cm–1m (6–12in–3ft 2in) for seed pods

POLYCULTURE POSITION

Glade

An iced radish, slathered in unsalted butter and dipped in salt, is the whole reason to learn to garden because you can only have this experience with a freshly pulled radish, preferably pulled within the hour. What sounds like a ridiculous indulgence is actually just a quick lunch if you've got radishes to thin.

Radishes are crazily quick to grow, but equally, if you want nice round, plump roots to eat that are not too mustardy, you need to be very quick at thinning otherwise you'll end up with something pithy and less delicious. Try and sow as thinly as possible as this solves this problem, and keep well watered. I often add radish seeds into other polyculture broadcast sowings, with turnips, lettuce, spring onions, rockets, dill and parsnips. Thin to 2–2.5cm (¾–1in) apart as early as possible.

For Asian radishes or mooli delay sowing till mid-summer otherwise they may bolt. Sow in the ground a little deeper than ordinary radishes as the seed is larger, 1cm (½in) deep in sunken drills 4cm (1½in) deep, pulling the soil around them as they grow otherwise you get green tops. Spacing depends a little on cultivar, but in a mixed polyculture they should be 20–25cm (8–10in) apart getting them established first and then broadcasting over them with lettuce, dill, carrots (short types work best), rocket, spring onions. You can also raise mooli in modules to plant out before the tap root hits the bottom and these can be planted into existing polycultures. Most cultivars are ready to harvest in eight weeks, but will stand mature for some time without deteriorating.

All radishes have edible seed pods, so if they do bolt, let a few flower and eat the young, green seed pods as they are as delicious as the roots. They also pickle well too.

Parsnips

Pastinaca sativa

POSITION
Sun to part shade
SOIL CONDITIONS
Do best in organic rich soils
FLOWERING PERIOD
Second year from mid-summer
HARVESTING PERIOD
Roots from late summer into winter
HEIGHT
50cm–2m (20in–6½ft) when in flower
POLYCULTURE POSITION
Glade

I am a big fan of being a lazy parsnip grower. Every three or so years I buy a packet of open-pollinated parsnip seed, broadcast sow them in an open spot around mid-spring when the soil is warm and wet, wait for them to come up, then thin to 20cm (8in) or so between plants. Often in the same space I will sow dill, lettuce, radishes, spring onions, rocket and even baby beet to harvest as baby leaves between the parsnips. Then I eat three quarters of these parsnips over the winter, cover the rest with straw if I think the soil will freeze, and allow these to flower the following year. If the soil is workable in late winter, I may move them before they come back into growth as parsnips are very tall in flower, around 1.5m (5ft), with a large umbel of acid yellow flowers that are so greatly loved by bees and hoverflies that the whole plant sings. I let the later-flowering plants self-seed, but not the earliest flowering ones as this seed would make early-flowering plants which tends to mean smaller roots.

Each plant has at least 500 seeds and you'll find that hundreds of them will come up all over the place. Weed out those that have chosen a ridiculous space and enjoy the rest at various sizes. Tiny parsnips are a bit of a revelation, need no peeling and are quite delicious. After two or three years you tend to lose the original vigour and need to buy in named seeds, but it's a very good way to grow parsnips in a lazy manner and enjoy their stunning flowerheads.

Carrots

Daucus carota subsp. *sativus*

POSITION
Sun
SOIL CONDITIONS
Do best in organic rich soils
FLOWERING PERIOD
Second year late summer onward
HARVESTING PERIOD
Roots from mid-summer
HEIGHT
30cm (12in)
POLYCULTURE POSITION
Glade

Truth be known I have a terrible relationship with carrots. Lovely as they are to eat, I am surrounded by banks of cow parsley from the train line that runs between my house and the allotment so I get waves of carrot fly. The serious need to cover all carrots to get a crop just doesn't work aesthetically for me so I tend to buy all my carrots in. There, I've confessed! However, I do sow carrots in pots by the back door; these are harvested young and sweet and this works well. The pots are 50cm (20in) across and deep and the carrots love them. I broadcast sow the carrots so that very little thinning is needed and cover the whole pot with Enviromesh to keep off carrot fly, then harvest handfuls of baby carrots as I need them and get my fix this way. There is some anecdotal evidence that sage, rosemary and onions help to deter carrot fly. This sadly has never worked for me, but it may for you.

However, you may not suffer from carrot fly, you lucky thing. Carrots are very slow to germinate, often taking over 14 days particularly if the soil is cold and damp, so try and sow when the soil is warm. It takes around six to eight weeks for baby carrots, and five months for a carrot to mature for storage so sow from mid-spring onward. Slugs love them and often you may think that your entire crop hasn't germinated, but if you get right down to soil level, you'll see tiny little stalks where your babies once were. Leaving large old cabbage or kale leaves on the bed and collecting the marauding slugs that will sleep underneath them in the daytime is one trick. A fine tilth is another, but ultimately, it's you versus the slugs and go at that however you wish, just don't use chemicals.

If you are sowing in a broadcast polyculture, then I wouldn't mix the carrots in with the other seeds. I would sow everything else and then sow the carrots singly in appropriate spaces. I've had some success in the past with sowing carrots between dwarf French beans, widely spaced, and among parsnip in the same manner. I've also very successfully grown carrots in large pots that had held first early potatoes as a secondary crop.

Annual Kales and Cabbages

Brassica olearacea

POSITION
Sun to part shade

SOIL CONDITIONS
Any good garden soil

FLOWERING PERIOD
Early spring to mid-summer

HARVESTING PERIOD
Leaves throughout the growing season

HEIGHT
30cm–1.2m (12in–3ft 10in)

POLYCULTURE POSITION
Glade

My passion for kales, and for that matter cabbages, runs deep, but I don't tend to grow many cabbages because they rule like kings and make too many demands for space. Kales, being more ancient types with wilder genes, are used to other bedfellows. You can plant right the up to the stem of a kale and it won't mind, but you try that with a cabbage and you'll find quite the opposite, plus cabbages have a habit of dropping their large lower leaves and squashing whatever is beneath.

If you want to grow cabbages I suggest raising them in pots for as long as possible before planting out well-established teenagers and giving them far wider spacing than you imagine – at least 50cm (20in) or more. I would probably stick to spring and autumn cabbages that tend to be smaller and more compact rather than gigantic summer sorts, but if I had lots of space, I would certainly make room for winter cabbages 'January King' and Savoy types with their distinct crinkled, emerald green leaves, but they're not great for a polyculture.

Annual kales are more tolerant and look so good in a mixed scheme, adding much needed height, then before they bow out they'll give you wonderful spring flowers. The dark, steely blue-green of Tuscan kales with a flurry of pale lemon yellow flowers in early spring is a really good sight, and they look handsome between the tulips and forget-me-nots and daffodils. These flowers are much loved by bees and flower over quite a long period. As the last of them starts to fade you can whip the whole thing out into the compost.

I sow kales mid- to late spring in seed trays, sometimes potting them up, sometimes transplanting them straight into the beds depending how well they are doing. Don't leave them sitting around in modules for too long. Any brassica that is not happy will start to turn purplish with stress and although they seem to tolerate a lot, if stressed at seedling stage they seem to hold on to the trauma for the rest of their lives and never do that well. Kales aren't hugely fussy but don't like very acid conditions and prefer free-draining soil. If you think your spot might

flood in winter, plant them on a mound with good staking if necessary.

They're a diverse bunch – dwarf ones grow no more than 30cm (12in) tall and need that sort of distance from others, larger types easily grow to 90cm (3ft) and can be spaced up to 75cm (2¼ft) apart. The denser the polyculture around the kales the wider they need to be spaced. I tend to dot them about, sometimes creating clusters for impact.

Once plants are mature, they need very little attention, an occasional liquid feed and perhaps covering in late winter against pigeons. If you stake tall perennial kales the pigeons will wise up to just sitting on top of the plant and devouring it, particularly in winter, at which point I sometimes make little tents of netting over them so I can carry on picking.

My favourite annual kales are 'Black Tuscan', 'Palm' kales of which 'Dazzling Blue' is my favourite, 'Sutherland', 'Red Russian', 'Siberian' and 'Red Ruffled'.

Frilly Mustards

Brassica hybrids

POSITION
Sun to part shade

SOIL CONDITIONS
Any, not fussy, can grow in both very acid and very alkaline soils

FLOWERING PERIOD
Mid- to late summer

HARVESTING PERIOD
Leaves mostly over winter and early spring

HEIGHT
30–50cm (12–20in)

POLYCULTURE POSITION
Understorey and glade

'Golden Frills' and 'Red Frills' are a cross between a kale and a mustard, large, annual plants around 30cm by 30cm (12in by 12in) with golden-green or red lacy leaves that are sweet and spicy, but not so hot that you can't eat them raw. They pair very well with slightly bitter greens and work nicely with robust vinaigrettes. They can also be briefly steamed or used in stir-fry. Sow direct or in modules in early spring or again in mid- and late summer for winter pickings. The flowers are edible too, but remember to leave a few to set seed.

Annual Leeks

Allium ampeloprasum

POSITION
Sun
SOIL CONDITIONS
Good garden soil that is free draining
FLOWERING PERIOD
Second year late summer onward
HARVESTING PERIOD
Early autumn through to winter
HEIGHT
60cm (2ft)
POLYCULTURE POSITION
Glade

I suffer terribly from allium leaf miner that makes lacy doilies of any leeks grown in the open. To keep them covered from the miniscule female fly you need to bury fine mesh netting around the leek bed and after many years of failing to be the sort of person that buries netting appropriately, I have largely given up on annual leeks to eat the wild, perennial leek, widely known as Babington leeks, although I do sometimes sow early leek varieties direct around mid-spring and harvest them when they are pencil thickness.

If you do want to grow leeks, sow them when the soil is consistently above 7°C (45°F) day and night, which may mean you wish to start your leek seedlings off early indoors at around 15°C (59°F) to transplant out. Sow 1–2cm (½–¾in) deep, make several sowings for successional harvests, sowing the last lot around mid- to late spring. Leek size is influenced by spacing. If you want leeks to stand over winter you need to thin them to be about 25cm (10in) in each direction, leaving enough spacing to grow lettuce or baby fennel in between. For slimmer leeks, sow in wide drills so the seed is 1cm (½in) apart, with 15cm (6in) between drills and harvest when the leeks are the size of a pencil. Mini leeks can be left standing for longer than you might expect without any loss of flavour, but the leaf miner might get them.

You can multi-sow leeks in groups of three, which looks very attractive, and gives you slimmer, sweet leeks in summer. Space these 25cm (10in) apart. Plant leeks into holes to get a blanched stem – make a hole 15–20cm (6–8in) deep with a dibber and drop a chopstick-thick leek in the hole, then water with a fine rose on the watering can so that the soil naturally settles around the hole and blanches the stem.

If you fail to harvest all your leeks by spring, leave a few to flower – they are very handsome alliums with a lovely light lilac flower on a drumstick head. They make a very tasty garnish or leave them for the bees and harvest the dry stems just before they set seed for flowering arranging. 'Musselburgh' is a very reliable open-pollinated variety, 'Lancelot' and 'Jolant' are suitable for early cropping for thin baby leeks.

Swiss Chard

Beta vulgaris subsp. *vulgaris*

POSITION *Sun to part shade* **SOIL CONDITIONS** *Any good garden soil*
FLOWERING PERIOD *Second year late spring onward* **HARVESTING PERIOD** *Through the growing season and into winter* **HEIGHT** *45–60cm (18in–2ft)* **POLYCULTURE POSITION** *Glade*

Swiss chard is such a good friend to the gardener. It is tough, long lasting, beautiful and so sustaining, just a few leaves go such a long way in the kitchen. That slightly metallic, earthy taste of its thick midribs and the soft, grassy green of the upper parts lend themselves so well to so many dishes. Swiss chard is a very close sibling to sea beet, the wild ancestor of beetroot, and has over the years been bred from red to white to purple, orange, pink and back to red again with ever thicker midribs and ever more ruched leaves. The most famous of these is the colourful mix 'Rainbow Chard' that comes in all the colours of the rainbow other than blue. It's a good mixture, but a cliché and you may only end up with one of each, so I grow single colours instead. 'Pink Passion', 'Golden Chard' and the immense 'Fordhook Giant' with the widest midribs of the lot in a lovely off white. I take immense pleasure in finding the best bedfellows for my chards, putting 'Golden Chard' with sunny, spidery yellow dahlias or 'Pink Passion' with pink-stemmed Chinese celery and *Geranium* 'Patricia', a sterile version of *Geranium pstilostem* that flowers all summer long, with a few *Monarda fistulosa* for a very rosy combination.

Swiss chard is quite hardy in my part of the world, but in the depths of winter plants tend to hunker down with rather small leaves, then grow luxurious with lengthening spring days before bolting to flower. If you are only growing one variety you can easily save your own seed. Swiss chard is wind pollinated, so as long as you have five or more plants relatively close together you'll find you get a lot of seed. They do take up space between flowering and setting seed, but I like the way they tower when in seed, adding an extra dimension to the early summer garden when other plants are generally quite short. If you have space, I suggest seed saving every three years to keep yourself in supply. There are two windows for sowing, in spring for crops through the summer and autumn, and again after mid-summer for plants that will overwinter and provide picking into spring. Don't sow in late spring or early summer as these plants will tend to bolt in late summer heat.

Once established, Swiss chard is very happy nestling among others, but if too overcrowded at the beginning of life it does tend to struggle to get going. In a multi-polyculture sowing grow it with lettuce, herbs, rocket and baby greens that you'll harvest quickly from around the emerging chard. I tend to module raise chard and plant it straight into an existing mix when it's about 15cm (6in).

Peas and Beans

Peas and beans are excellent crops for they feed you and feed the soil with their nitrogen-fixing roots, which have a symbiotic relationship with soil bacteria, rhizobia, that can take atmospheric nitrogen and turn it into a form that is available for the plants.

In this way they are a key group for helping to maintain soil fertility. I've several times grown a crop of peas and beans on poles over the ocas. The ocas don't seem to mind the shade one bit and thrive with all the nitrogen the beans bring. I wouldn't do it year in year out, but I was surprised by what happy bedfellows they are. If you want to crop in between or in the centre of the trellis, choose a quick-growing cut and come again lettuce mix that will be harvested before the beans shade everything out or space each bean station upright as much as 40cm (16in) apart to allow light in.

I use dwarf peas a lot in the main garden as a quick crop of sweet tendrils and pea tops to pick for salads, sowing direct and harvesting as cut and come again, allowing them to flower and then removing them by mid-summer for something else to go in their place. Dwarf peas look fantastic growing in wavy drills around the tulips and help to hide the foliage as it is dying back. Dwarf peas are also an excellent solution to any gaps that appear in the garden. I use a Joy Larkcom trick of using a large, round glass jar (the sort pickles come in) to create a depression, fill it liberally with pea seeds and cover. This way you can create nice dabs of green here and there. If you choose a purple-leafed radish for microgreens you can alternate dabs of vibrant green with purple and it looks rather marvellous. Finally, rather than buy seeds in do let some of your dwarf peas flower and pod so you can collect your own seed. As you haven't self-isolated the flowers, they may have cross pollinated if you or your neighbour are growing different varieties, but they will be fine for pea shoots.

Runner Beans

Phaseolus coccineus

POSITION
Sun

SOIL CONDITIONS
Organic rich soils

FLOWERING PERIOD
Mid-summer onward

HARVESTING PERIOD
Summer to early autumn

HEIGHT
2m+ (6½ft+)

POLYCULTURE POSITION
Glade

I grow a lot of runners as much for the fat seeds that I dry for winter use as for fresh beans. For this reason, I tend to grow either white or black bean varieties. I have a number of favourites. 'Moonlight' is a modern runner bred for drought resistance with lovely white flowers and white beans. Young beans have a very good flavour. 'Painted Lady' is a 19th-century variety with bicolour flowers and excellent prolific beans of mottled purple. These tend to turn brown when cooked, but have a very good flavour. 'White Lady' is a selected pure white-flowering form with lovely white beans that make excellent stews. 'White Emergo' is another white-flowering, white bean that is particularly good in casseroles and can be used as a butter bean substitute.

'Jackpot' is a dwarf form that works well in pots. I don't grow it in the ground as the beans are large and end up sitting on the soil, which inevitably means being munched by slugs.

'Black Coat' is from the 1600s and produces very short, fat pods that only have two or three beans, but these beans when mature have very shiny black coats and are excellent for stews. It's not an easy variety to find, so if you do, I suggest growing this variety on its own for a year so you can save your own bean seeds.

French Beans and Shelling/Drying Beans

Phaseolus vulgaris

POSITION *Sun* **SOIL CONDITIONS** *Organic rich soils*
FLOWERING PERIOD *Mid-summer onward* **HARVESTING PERIOD** *Summer into early autumn*
HEIGHT *30cm–2m+ (12in–6½ft+)* **POLYCULTURE POSITION** *Glade*

I grow tall and dwarf varieties of all beans. The tall beans are such fun to play around with, creating wigwams and trellis and archways for them to grow up, but dwarf beans often end up being my mainstay particularly for drying and shelling beans for winter crops. I make two main sowings of both tall and dwarf, one indoors in mid-spring to have hardened off and planted out by late spring, and a second main sowing at the end of spring to cover any losses and also to give a succession to harvest. I often make a third and fourth sowing of dwarf French beans in early summer and again in mid-summer, cloching these last beans in early autumn to get one final harvest of young, tender fresh beans before I settle to a winter of dried beans. I have in the past tried shelling beans as late as early summer, but if the weather is anything less than perfect and turns wet in late summer, they have rarely fattened up and dried in time for autumn harvest.

If I'm sowing indoors I sow in the largest modules I have. If I'm sowing direct I sow two seeds per station, roughly 15cm (6in) apart for dwarf and 25cm (10in) apart for tall varieties, and thin to the strongest one. Beans and peas do best in fertile soils that don't dry out. The traditional runner bean trench that is filled with semi-rotting compost in early spring before planting is one such method, but being considerably lazier I just dib the seed into the ground and cover with a good layer of homemade compost. It can be quite rough – the worms will do the rest and the compost will act as moisture-retaining mulch.

'Beefy Resilient Grex', bred by the famous Carol Deppe of *Breed Your Own Vegetable Varieties* (a bible for seed savers everywhere), is a wonderful dwarf variety with black, brown and mottled beans in earth tones that are thin skinned and meaty with a good texture. They are truly good beans, adapted to a wide range of environmental conditions and cropping even in a short season. They've even cropped well for me in beds that have considerable shade.

'Black Coco' is another bush form, a French heirloom with delightful dark lilac flowers and shiny, round black beans that are excellent for stews and bean dishes.

'Cosse Violette' is a very vigorous, early, tall French bean with stringless purple pods and good beany flavour. 'Lazy Housewife' is a tall French bean with fat, buttery beans in the pod. It's late maturing and best for shelling. 'Cherokee Trail of Tears' is a tall bean with rounded red-green pods with a particularly good flavour. The beans are pretty too. 'Faraday' is the best dwarf I've found and can be sown as late as the beginning of mid-summer and still crop before autumn – stringless, fine green beans with a good snap.

Peas

Pisum sativum

POSITION
Sun
SOIL CONDITIONS
Organic rich soils
FLOWERING PERIOD
From late spring onward
HARVESTING PERIOD
Through summer
HEIGHT
15cm–2m (6in–6½ft)
POLYCULTURE POSITION
Glade

If you are limited on space then grow mangetout for the longest harvest as the more you pick the more you'll force the plants to flower to try to produce seed. I like the old French variety 'Ca-rouby de Maussane' with large, flat edible pods, and the lower-growing (around 60cm/2ft high) 'Sugar Dwarf Sweet Green'. 'Petit Pois Provencal' is speedy with sweet, tiny round peas perfect with feta and salad for early summer. 'Half pint' is the smallest dwarf I've ever grown at just 15cm (6in) high; small yields, but perfect for a windowbox or mixed in between early sowings of lettuce.

All peas need support of some sort otherwise the pods will sit on the ground and get munched. These days I tend to use old bed springs (cities have an endless pavement supply of these once you get your eye in) between strong supports. They are perfect for peas, offering plenty to hold onto, and because they are wide you can sow the peas under them to clamber through and they're even quite good at keeping pigeons off. I've yet to find a way to integrate peas into a true polyculture and grow them on the margins of beds. I also have never grown enough peas to freeze, though I do grow some for drying for soups but I concentrate growing peas for pea shoots and flowers for salads. Many good dwarf forms are happy to grow in a mixed polyculture and work well between beets, lettuce and low-growing herbs or in wide drills between leeks or garlic.

'Snow pea Usui' has been bred for a long picking season for pea shoots and the regrowth is tender and sweet. From China, it's well worth saving seed from as it's not always easy to get hold of. 'Parsley pea' has strange, parsley-like side shoots that are considered a delicacy. It's a self-supporting dwarf and works very well in pots but I've found it won't do well with any kind of shading so give it a sunny spot.

Cucurbits

The sight of winter squash laid down for storage is one of the great satisfactions of growing your own. I could dedicate a whole book to the subject of best-tasting squash, but I will say this, the summer squash are by far more interesting than the average courgette or zucchini.

My young plants (when they have four or more true leaves) go in a hole filled with well-rotted manure. When they are young enough that I can still get between the vines, I mulch with grass clippings or homemade compost and use a stick or bamboo cane to mark the centre of each plant so that I know where to water once the leaves get growing. I also twirl the vines around the plant, pegging them down with wire pins to make the most of space. Some years I train some up pergolas or wigwams.

When the plants are young, I use old window frames to make temporary cold frames around them, particularly if the summer is being slow. Early heat is everything. Once they get going, I remove the frames and the minute I have flowers I start feeding with comfrey tea every two weeks. I can't emphasize enough that if you want lots of fruit for winter storage you need to champion good growth.

Eat What You Grow

Winter Squashes

Cucurbita species

POSITION
Sun to part shade

SOIL CONDITIONS
Organic rich soils

FLOWERING PERIOD
Mid-summer onward

HARVESTING PERIOD
Early to mid-autumn

HEIGHT
1m+ (3ft 2in+)

POLYCULTURE POSITION
Glade

All winter squashes are best cured before they are stored. To cure the squash keep it somewhere around 20–25°C (68–77°F) for two weeks. This converts some of the carbohydrates into sugars and makes sure the skin is completely dry. After that they need to be stored at 12–15°C (54–59°F), any lower and they will rot, any higher and they will dry out. An unheated bedroom or hallway is usually fine, a shed is often too cold at night and too hot during the day.

There are so many winter squash varieties to choose from. The following are my biases, not 'Crown Prince' or 'Uchi Kuri' but they could easily be on the list as could hundreds of others. The only butternut squash I will ever grow again is Dan Barber's 'Squash 898' that grows no bigger than your hand and truly tastes of something; everything else I consider watery rubbish. I told you I was biased. I choose those that do well on my Midlands clay so I'm sticking with them, but I genuinely encourage you to experiment to find your best fit.

'Burgess Vine Buttercup' is a great choice for smaller gardens as it's not a huge vine and one plant can feed two people. With excellent, dense orange flesh and a superb flavour, if I were

to choose just one, it would be this. 'Hokkaido' is originally from Japan, but has been selected over the years for European climates. Coming from the Hubbard family of squashes it is very easy-going, always producing lots of fruit early whatever the summer holds.

'Thelma Saunders Sweet Potato' is a wild, rampant thing, but I love its deeply lobed leaves and it works very well grown up a strong trellis. It is quite late to set fruit, but once it does it really gets going, hugely prolific with many acorn-shaped, deeply ridged sweet squashes that store particularly well. 'Sibley Squash' is very early ripening, the grey, pear-shaped squash keep well overwinter. The longer you store them the sweeter they get.

I have a big thing for spaghetti squash which are generally seen as summer fruiting but they really don't taste that great if picked young – it's so much better to wait until the skin has hardened then the inside becomes something else. I grow a huge number and genuinely do use them as a spaghetti substitute over the winter. They need space as vines are rambling and vigorous, and need plenty of moisture to provide lots of fruit.

Courgettes and Summer Squashes

Cucurbita pepo

POSITION *Sun to part shade* **SOIL CONDITIONS** *Organic rich soils*
FLOWERING PERIOD *Mid-summer onward* **HARVESTING PERIOD** *Mid-summer to early autumn*
HEIGHT *1m+ (3ft 2in+)* **POLYCULTURE POSITION** *Glade*

I love crookneck, straight neck and scallop squash and particularly 'Tromboncino' and 'Tromba d'Albenga' because you can get them to climb into trees which makes them very useful for a small garden. You want to eat them when they are 30cm (12in) or so long when the skin has yet to harden. At that point, unlike so many watery courgettes, the flesh is firm and sweet. If the skin hardens the flesh starts to deteriorate inside and they are less lovely. They look truly strange and wonderful growing up through trees and trellises.

And I have a soft spot for tiny pattypans in yellows and pale creams and always grow one or two plants for a supply of these. I sometimes grow the round, green ball courgettes for stuffing, though I think a pattypan is as good, but long ago gave up on straight green courgettes. If I was to go back on this it would be for the classic striped Italian courgette, 'Romanesco', which is a good choice for nutty flavoured, firm green courgettes – when I think about them finely shaved raw, dressed with olive oil and lemon juice and salt, I think I might change my mind. But space is always an issue and I'd rather have a swan-necked oddity scrambling over something than a bush of endless straight green courgettes.

'Pattison Blanc' is an old white French pattypan with particularly dense flesh and a nutty flavour. 'Benning's Green Tint' is a US heirloom variety from the 1900s that is particularly attractive with highly scalloped, light green pattypan-shaped fruit. However, it is quite a large, vigorous bush, so not ideal in smaller gardens.

'Early Prolific Straightneck' has bright yellow warty fruit, best picked young when the skin is thin. The flesh is excellent, dense and cooks very well. 'Summer Crookneck' is a bush-forming summer squash with warty, bright yellow, bent-necked fruit – a good choice for smaller gardens as the bush remains very upright leaving space for growing around it. 'Rugosa Fruliana' is a variety from the Veneto region in Italy and has light yellow, warty fruit. It has a particularly good flavour to the flesh and is very prolific once it gets going.

'Tromboncino/Tromba d'Albenga' is a very vigorous vine that will find anything it can to climb; mine grow up into my apple and quince trees. I raise the young plants in 5-litre (9-pint) pots and when the roots start coming out of the drainage holes, I plant them on the edge of the tree canopy, training them to grow up the trunk and into the branches. They look magnificent and always manage to somehow reach the fruit. I pick them when they are no more than 60cm (2ft) long. I love the dense flesh and one plant will easily keep two people in more than enough fruit. These are quite slow to get going at the beginning of the season so it's worth growing a pattypan to fill that gap, but by early autumn you're rich in fruit. They also roast very well for freezing for winter storage.

'Serpente de Sicilia' is a climbing crookneck squash, similar to 'Tromboncino', but likes to bend even more dramatically, like a snake. It is quite sweet and used along with lemon rind to make a jam in Sicily.

Achocha is a curcubit from South America with a rampant habit (it could easily climb over a garden shed) and tiny, curious, pointed green fruit with soft spines. The young fruit can be eaten just like cucumber, sliced or added whole to salads. The older fruit can be cooked just like green peppers to which they taste remarkably similar. Just remove the black seeds first as they are unpleasantly hard, saving a few for next year's crop. 'Fat Baby' is the best variety for outdoor growing.

Salads

Salad plants are way more than just lettuce. I grow around a hundred plants that could happily leap into a salad bowl and make a fine meal and most of those are scattered elsewhere through this book. So this bit is a detailed look at the base ingredients for a good salad, the leafy sorts like lettuce, mustards and Asian vegetables that are some of the most productive plants for small gardens.

It takes a bit of energy raising them from seed in modules, but once established they can be picked for many months, look great and make a very big difference to the environmental impact of your food shopping. Salads nearly always need to be refrigerated to be transported any distance, those mixed bags of lettuce have to be washed, prepared and packed and this all adds up. Only the most robust lettuce varieties make it to a commercial system, whereas you have the luxury of growing the most flavoursome as all you have to do is walk them from your plot to your plate. Or in my case treat eating salad as a grazing experiment. I often eat my lunch on the go in the garden as I weed or work.

Lettuces

Lactuca sativa

POSITION *Sun to part shade* **SOIL CONDITIONS** *Any good garden soil*
FLOWERING PERIOD *Mid-summer onward* **HARVESTING PERIOD** *Year round*
HEIGHT *30cm–1m (12in–3ft 2in) in flower* **POLYCULTURE POSITION** *Glade*

Lettuces are broken down into two broad categories: they either heart up or they don't, though there are a few that want to keep a foot in both camps. Hearting types include cos (called romaine in the US) and the flat or 'cabbage' type, which are subdivided into butterheads or crispheads. The non-hearting types include loose-leaf and stem lettuces.

Cos are large, very upright plants with crisp, thick, distinctly flavoured leaves, perfect for a Caesar salad, a burger or a simply dressed salad. They are known to stand better in hot, dry conditions and there are some that are very cold hardy for winter production. 'Winter Density' is rightly well known. Semi-cos are smaller types, the most famous being 'Little Gem' with its delightfully sweet, crisp leaves.

Butterheads are soft cabbage-type lettuces with flat, rounded leaves that wilt very quickly after picking, so aren't seen much commercially. They are best for the shorter days of late summer, autumn and early winter. Varieties include 'Tom Thumb', 'Plenty', 'Arctic King' and 'Valdor'. Crispheads are also cabbage types but with crisp, textured leaves of which the most famous is 'Iceberg'. The commercial sort has been bred to stand well in hot weather without bolting (something a butterhead will never do) and keep well after picking, all at the expense of flavour but they are often tasteless. Exceptions to the rule are the heritage Batavian sorts including 'Rouge Grenobloise' that are good for winter growing.

Loose-leaf types are best for long picking production because without a heart you can pick individual leaves as you require. My favourites are the oakleaf varieties that have deeply lobed leaves and the 'Lollo' types which are deeply curled. They have very tender leaves that are easily damaged by hail, wind and rain and wilt like wet handkerchiefs soon after picking, so do so at the last minute. Varieties include 'Mascara', 'Freckles', 'Lollo Rosso', 'Salad Bowl' (green and red), 'Frisby', 'Catalogna' (oakleaf).

Stem lettuces are Asiatic lettuces that are sometimes also known as Celtuce. If you are the sort who loves the bitter base of a 'Little Gem' or 'Romaine' then this plant is for you. They need a good summer and lots of moisture to do well.

For space, efficiency and best use of seeds, I start nearly all my lettuce off in seed trays and then prick out into modules. In early spring I raise all my seed indoors to transplant to modules outdoors. By late spring this is all happening outside. Lettuces are best transplanted at the four or five true leaf stage. I plant 15cm (6in) apart for small plants, 25cm (10in) for large butterheads, Batavian and large cos. Lettuce germinate at quite low temperatures and very poorly at temperatures over 25°C (77°F) so if you are sowing at peak summer temperatures you need to sow in early afternoon so that the most critical period of germination (a couple of hours after sowing) is in the cooler evening temperatures.

Lettuce can easily become bitter, bolt early or get diseased but this is nearly always because of slow growth due to lack of water. This is one crop to prioritize if conditions are dry, watering in early evening or early morning. Bitter, tough leaves are particularly a problem in containers so make sure that you are watering deeply, until you see it coming out of the base of the pot as this indicates the water has moved all the way through the pot and not just the first few inches.

I sow my lettuces in intervals from early spring to early summer and then again from mid-summer into the first two weeks of early autumn. This keeps me in lettuce year round. To avoid gluts and gaps make each successional sowing when the seedlings of the first sowing are up; this is usually about two weeks.

One of the best ways to harvest lettuce is to not harvest the whole head, but to remove the outer leaves. Discard any that are yellow or too munched and just remove up to three leaves per plant and move onto the next one until you have enough. Harvested this way the plant will continue to grow and can stand for up to three months. The stem will start to become exposed; don't worry about this, it is in your favour as the leaves will no longer sit on the soil and there will be less places for slugs to hide. If possible always try and spot water lettuce rather than watering a whole row – if the soil between plants is kept dry, you'll find you get far less slug damage. Finally, there seems to be strong anecdotal evidence that red lettuce is more slug proof than green.

Pak Choi

Brassica rapa var. *chinensis*

POSITION *Sun to part shade* **SOIL CONDITIONS** *Organic rich soils*
FLOWERING PERIOD *Late spring to summer* **HARVESTING PERIOD** *Year round*
HEIGHT *50cm (20in)* **POLYCULTURE POSITION** *Glade*

This is one of China's oldest vegetables, in cultivation since the 5th century. The familiar fat, white butts of supermarket pak choi are not easy to achieve at home as those broad bases require a lot of watering and even warmth. The joy of this plant is that all parts are edible and it can be used at any stage, so there are lots of other ways to enjoy it, like drying the large mature leaves for winter use. There are also many different kinds: the white-stemmed type that you find in the supermarket, soup-spoon types with smaller leaf parts, types with green leaf stalks and rosette-forming pak choi that sits low to the ground.

The ideal temperature for pak choi is 15–20°C (59–68°F), so they tend to be best sown from mid-summer onward and grow quickly to bulk up over autumn. However late summers can mean that temperatures soar and these plants bolt if it's too hot. This is equally true if you sow pak choi in spring as the summer heat can mean premature bolting. Green-stemmed varieties are the hardiest. If you have a greenhouse raise plants in there with plenty of water and good ventilation. Space 15cm (6in) each way for small plants, 45cm (18in) each way for large plants or enjoy them at seedling stage, in which case you can sow pak choi at almost any point in the summer for a quick-catch crop of tasty small leaves. It's an ideal one for between caulis or other large brassicas or beneath tall bean poles. There's a Chinese trick of growing carrots and pak choi together with a ratio of two parts carrots to one part pak choi and I've found, in the past, this can be quite effective at confusing the carrot fly.

Rosette Pak Choi
Brassica chinensis var. *narinosa*

This grows in large, flat rosettes and is used just like ordinary pak choi. All parts are edible including the young flowering shoot. The flavour is stronger than ordinary pak choi, but I think superior, and it's a much tougher plant, less demanding of perfect conditions and rather slug proof. Over the years I've rather given up on ordinary pak choi in favour of this one. It also makes excellent ground cover as the large, flat heads suppress weeds and it can be used as a decorative bed edge. Sow mid-summer for outdoor crops and late summer into early autumn for one that will be need to be covered. This makes an excellent crop for a cold greenhouse or polytunnel, but it's not so great for pots because you end up with one plant in a pot. The more space you give it, the larger the plant will grow. For small rosettes leave 15cm (6in) between plants, for large ones 30–40cm (12–16in).

Choy Sum
Brassica rapa var. *parachinensis*

The flower shoots of nearly any brassica can be eaten when young and tender, treated just like sprouting broccoli, and choy sum is grown specifically for this reason. Like sprouting broccoli, when the main head has been cut you get numerous smaller edible side shoots to harvest. Choy sum is not frost hardy and needs high summer temperatures to do well, so the best sowing seems to be from mid- to late summer with hope for an Indian summer. It's one to follow on from potatoes, garlic, peas and onions. If you aren't fussed about getting big plants you can grow it very successfully in a polyculture mix. Try it with a final sowing of dwarf French beans in mid-summer, perhaps with some peas for shoots, radishes and aim to thin so that the plants are 10–15cm (4–6in) apart and pick just as the first one or two flowers are opening, but the rest of the buds are tight for tender stems. Harvest the main stem when 10–15cm (4–6in) long, snapping it off, but leaving the plant intact. Side shoots should appear in the following weeks. Or harvest the whole plant young when the first flower shoots appear.

All of these brassicas can suffer from flea beetle, a small, shiny black beetle that eats tiny holes in the leaves. The best solution is to grow them in a patch where you can cover the whole thing with fine mesh netting, which needs to be done immediately after sowing because if you wait till the plants are up and cover them then, you'll just trap the flea beetle in its idea of heaven!

Mizuna

Brassica rapa var. *nipposinica*

POSITION *Sun to part shade* **SOIL CONDITIONS** *Organic rich soils*
FLOWERING PERIOD *Late spring to summer* **HARVESTING PERIOD** *Year round*
HEIGHT *50cm (20in)* **POLYCULTURE POSITION** *Glade*

This is one of the prettiest Asian vegetables. It originates from China, but has been cultivated since antiquity in Japan, where it is much loved and prized. It can be used at any stage from seedlings to flower shoots. Mild flavoured like all of this group it makes an excellent addition to salads, but works equally well cooked. It is also one of the most bolt-resistant of this group, so you can sow year round but in practice it is best sown from early spring to early autumn outside, with a final sowing in late autumn under cover. Sow direct or in trays or modules and transplant after 2–3 weeks, making this an invaluable fast green for the hungry gap. For small plants space 10cm (4in) apart, 20–25cm (8–10in) apart for medium-sized plants and if you want a full head, 30–35cm (12–14in). It is an excellent polyculture plant as it is good at intercropping between other slower-growing brassicas, kales and cabbages or between corn or courgettes. Because you can eat it at any stage, you can sow it quite thickly, eventually thinning so that a few mature plants are left. I tend to use it most in late summer and early autumn mixes and often create an Asian cutting patch, mixing it among kales that will stand over winter with mibuna, choy sum and mustards. It appears in many stir-fry seed mixes for this reason.

Mustards

Brassica juncea

POSITION

Sun to part shade

SOIL CONDITIONS

Organic rich soils

FLOWERING PERIOD

Spring to summer

HARVESTING PERIOD

Autumn to late spring

HEIGHT

Up to 50cm (20in)

POLYCULTURE POSITION

Glade

Mustards are hot, hot, hot and rugged. They are best grown in autumn and winter, allowing them to flower in spring, as the flowering stems are edible too. If sown in spring, they will bolt immediately to flower; pretty, but pointless as all the spiciness intensifies to wasabi-like heat which is frankly not pleasant. Mustards are slower growing than many of the other Asian vegetables and should be mainly sown in mid-summer for autumn and early winter production. They will need some form of protection if grown outside over winter or else the leaves will become too spicy. Space 20–40cm (8–16in) apart depending on the variety but don't go overboard with mustards, you won't use that many.

They can be broken in to purple-leaved forms like 'Red Giant' and 'Osaka Purple', which has deep purples and reds as the temperature drops in autumn, and green-leaved forms, such as 'Green Wave', which is very cold hardy and can be grown outside all winter. 'Dragon's Tongue' is best in a polytunnel for late winter with lovely wiggly, white midribs and good flavour, and 'Nine Headed Bird' has an excellent flavour and texture, again better in a polytunnel for winter greens.

All mustards should be eaten young for salads, large leaves tend to be too mustardy but can be cooked in stir-fries or added to soup or rice. Mustard oils become bitter with too much heat, so cook on high heat for no more than 30 seconds, adding the leaves as the last ingredient to any dish.

'Golden', 'Red' and 'Green Frills' are three forms of kale-mustard hybrids that combine the best of both worlds. You get a very tough, very slug-resistant, not too hot, very frilly, salad for all seasons. This is one of those that I let self-seed around if I have space and patience to leave them to flower and set seed. It's an invaluable salad ingredient that looks beautiful both on the plant and in the plot.

Herbs

Herbs are very important in my garden. I use them for cooking, but medicine too and the truth is this could be a very long list, but here are few that I couldn't do without and are strong and robust enough to enjoy the muddle of polyculture.

Perilla or Shiso

Perilla frutescens

POSITION *Sun* **SOIL CONDITIONS** *Any good garden soil* **FLOWERING PERIOD** *Mid- to late summer* **HARVESTING PERIOD** *Leaf in early spring before flowering, young flower spikes are also edible but save many for seed in early to mid-autumn* **HEIGHT** *45cm (18in)* **POLYCULTURE POSITION** *Glade*

Perilla is a Japanese herb that has a very distinct flavour. There's a hint of basil, I think curry or perhaps cinnamon and perfume, with notes of coriander and citrus. It is widely used in Japanese cuisine as a garnish, as salad, in tempura with rice and infused in vinegar. The purple-leaved forms are used to dye the famous ume plums. The seedling leaves are particularly good in salads as a microgreen, the seed is used a condiment and often preserved in salt and used in pickles, tempura and miso. The seed of the purple-leaved form is favoured for shichimi, a spicy-salty condiment particularly good on rice.

Perilla is best treated like basil, but it can be hardy outside and I grow a mixture of purple- and green-leaved forms partly just because the frilly leaves look so wonderful in the garden. It first came to the UK as a Victorian bedding plant and it makes for some very fine edging. It is very fast growing and in flower will reach to about 45cm (18in) tall and 30cm (12in) wide. It certainly thrives best in sun in well-drained soil, the green form can suffer from slug damage if grown in too much shade. Seed germinates best if it's fresh, no more than a year old and should be soaked overnight before sowing. Germinate on heat, at around 21–24°C (70–75°F) and grow on around 16–18°C (61–64°F), hardening off before planting out well after the last frost. It will not tolerate too many days below 15°C (59°F) so you may wish to cloche the plants until they get going. Perilla can be grown very successfully in a pot or container and makes an excellent microgreen for windowsills.

Summer Savory

Satureja hortensis

POSITION *Sun* SOIL CONDITIONS *Any good, well-drained garden soil*
FLOWERING PERIOD *Mid- to late summer* HARVESTING PERIOD *While in leaf, before plant comes into flower*
HEIGHT *30cm (12in)* POLYCULTURE POSITION *Glade*

The leaves of summer savory are spicy, tangy with a marjoram-like flavour with a strong dose of pepper. They are very addictive and when added to salt can be used fresh or dried as a condiment for many dishes. Leaves are particularly good with beans, as they are said to reduce flatulence, but also with any tomato-based dish. Unlike winter savory, summer savory is an annual and in my experience doesn't self-seed. I start mine in trays indoors to plant out when the last frost has passed. It won't tolerate shade or overcrowding from other plants, so needs to be grown near the path or bed edge, which also helps with picking. You can harvest leaves as you need, but it's so good that its worth growing more than you might think and harvesting whole plants before flowering for drying for winter use. You can follow on in the same space with winter greens such as rocket, mustards and rosette-forming pak choi.

Papalo and Quillquina

Porophyllym coloratum and *Porophyllym ruderale*

POSITION *Sun* SOIL CONDITIONS *Any good, well-drained garden soil* FLOWERING PERIOD *Mid- to late summer*
HARVESTING PERIOD *Late summer* HEIGHT *30cm–1m (12in–3ft 2in)* POLYCULTURE POSITION *Glade*

These two are both from South America, quick-growing annuals with intense scent and citrus-spicy flavour. Both are ideal for salsa, chillis and refried beans, tacos or with avocado. Quillquina has pointy leaves with a slight blue-green tint that grows to around waist height. Papalo has rounded, slightly scalloped blue-grey leaves and again grows to waist height. Both need well-drained soil and a sunny position. They won't tolerate shade, particularly when young, or they tend to get slug damaged. Rather than let this happen I sometimes grow them in large, 30cm (12in) pots on the patio till they are several feet high and then plant them into the polyculture. Papalo looks very good with other silvery, grey-blue leaves like lavenders and achilleas. I think both look best grown in small swathes of three or five plants together rather than dotted about. Sow indoors on heat late in spring and keep the plants warm in a cold frame until you can plant out in early summer as any cold nights will shock growth.

Tender Things

There are a handful or so of delicious things – tomatoes, chillies, cucumbers, tender herbs and citrus fruit – that need to bake in warm conditions for the best flavour and yields, and sulk if they get too wet and cold. Rather than make them shiver in less than perfect summers, it is easier to just turn up the heat for them. Cold frames, greenhouses, polytunnels, high cloches, hotbeds, porches, conservatories or just some strange contraption that you've fashioned out of junk can offer a warm room that will make all the difference with harvests and keep certain diseases at bay.

For many tender crops the diurnal day/night temperature needs to be as even as possible to get strong, fast growth. Where nights dip below 10°C (50°F) and days are 15–20°C (59–68°F), protected cropping will even out this drop. Clearly, though, once you put a cover over a plant you keep off the rain, which means you need to water and, if the plants are growing in pots, feed too. You should keep water off the leaves of tomatoes when blight is around (blight is airborne and damp leaves help spread it about). Chillies dislike having wet feet, as do basil plants at night. Cukes need to be kept damp for fat fruits.

I have a small polytunnel on the allotment, where I mostly grow cucumbers. It doesn't get great light, so tomatoes tend to ripen very late, but cucumbers seem unfazed by the shade from trees. Beneath the cucumbers I grow salads and herbs, mostly dill because I am near obsessed with fermenting cucumbers into sour dill pickles. Someone wisely once said to me that if you are going to put up a polytunnel put up the largest one you can afford, as once you're growing it will never be big enough and oh boy do I wish I had listened. You don't have to buy brand new – eBay and local pages often have hoops going cheap and you can buy new plastic.

Allow for more time than you'd imagine for erecting a tunnel and remember that fierce winter winds will lift up the whole thing if it doesn't have properly secured feet. At home I have a wooden patio greenhouse. Tiny by many standards, it seems princely to me and allows me to raise seedlings in spring and then take out the shelves for pots of tomatoes and chillies in summer.

Eight tomato plants and as many again of chillies keep me in enough frozen passata and dried fruits to make it through to early spring each year. In winter I turn the tunnel over to a steady supply of mustards and greens that our mild climate allows all year round.

Tomatoes

Lycopersicon species

POSITION
Sun

SOIL CONDITIONS
Organic, rich soils

FLOWERING PERIOD
Mid-summer onward

HARVESTING PERIOD
Summer to early autumn

HEIGHT
30cm–2m (12in–6½ft)

Tomatoes are tropical plants, they hail from South America where their ancestors scrabbled over rocky ground and clambered up anything that they could. The original wild tomatoes are tiny, like small currants, and over millennia selection has taken those brilliant, delicious morsels and turned them into round, fat fruit. You have to honour a plant's origins and essential nature in damper, milder climates by starting off the toms indoors with heat and only taking them out to the wider world when the weather warms up.

You can buy young plants to transplant and this is a good idea if you are a beginner, but if you have a sunny windowsill, a cheap heated propagator and the will, you can grow any toms you wish, from the bog-standard 'Gardener's Delight' to some rare and delicious heirloom variety. If you have a less than sunny windowsill then you can up your game with fluorescent lights or grow lamps to increase the daylight. You want to keep the lights just above the plants as there's an inverse square relationship between light intensity and distance – at twice the distance you get a quarter of the light intensity.

Tomato seedlings germinate best at 24–30°C (75–86°F), but once they are up, they need to be off that heat as quickly as possible. The seedlings grow best at 16–21°C (61–70°F) and the best growth happens when their feet are around 16°C (61°F) and their heads are around 21°C (70°F). I sit mine on the kitchen floor beneath grow lamps and near the radiator, and this makes for very happy toms as long as I keep on top of watering. The last thing you want is long, leggy growth, as that weakness will be carried through the plant for the rest of its short, but brilliant life.

Once your seedlings are up, pot them on into bigger pots (I tend to use square 9cm/3½in pots). Never handle any seedling by its stem, as it's too easy to damage; use the seedling leaves. Hardening off takes around two to three weeks and is best achieved in a cold frame for plants that are going outside. If your toms are just going into a cold greenhouse you may need to offer them some fleece for the first couple of nights. What's perhaps more important is that you don't just turf them out on a very sunny day as the difference between indoor conditions and outdoor ones is huge and a bright sunny day will cause the plants stress and they may wilt as they struggle to adjust.

The larger the root run you can give your tomatoes the happier they will be, the less watering you will have to do and the stronger they will grow. How big the pot depends a little on the variety, there are ultra-dwarfs that will crop in 30cm (12in) pots and indeterminate (vine tomatoes as opposed to determinate, bush tomatoes) that won't be happy in anything less than 30 litres (7 gallons). When transplanting, remove the lower two sets of leaves and plant up to the next set to bury the stem as this will cause the stem hairs to turn into roots (magic!). It's best if you can place the transplant in at a slant as this will help establish a vigorous secondary root system.

You will need to support your tomato if it is indeterminate (vine) and you will either need a very long pole, a cage or string. If you are growing a small determinate variety (bush) then you may just be able to get away with a few pea sticks.

You may want to prune or pinch out the side shoots of indeterminate varieties but you don't need to prune determinate varieties as they naturally grow into a bush shape.

There is some evidence that pruning out the side shoots of vine tomatoes gives you earlier, bigger fruit, and early fruiting in British short summers is important. I have come to a point where I prune out the lower side shoots till they hit about waist height and then I let the plant do what it likes because the higher side shoots will fruit. The truth about tomato growing is that there are many ways of going about it and all of them are right for someone.

However you will need to water consistently as thirsty toms produce very small fruits that will crack or get blossom end drop – when the bottom of the fruit turns black from inconsistent watering. When the first flowers appear start feeding in earnest. I do this every week with a seaweed and comfrey feed.

There are hundreds upon hundreds of tomato varieties. I have a few favourites.

'Matt's Wild Cherry' is the closest we gardeners can get to a truly wild tomato. The cherries are tiny, but so delicious. It's the sort of thing you let grow completely wild in some baked but less than desirable spot in the garden and nibble on as you weed. I often grow mine up through an old rambling rose. It shows good resistant to blight.

I grow 'Black Cherry' because I love the taste of those dark, round cherry toms for salad, but I could argue for 'Yellow Pear', another work horse, and 'Gardener's Delight' because it's open pollinated and quite delightfully reliable and tasty. 'Legend' is also blight resistant and produces round, bright red tomatoes. This one can be grown outside if necessary and will still fruit. If I've run out of space in the patio greenhouse, I grow this in a pot on the sheltered part of the patio.

Chillies

Capsicum species

POSITION

Sun

SOIL CONDITIONS

Do best in organic, rich soils

FLOWERING PERIOD

Mid-summer onward

HARVESTING PERIOD

Late summer to early autumn

HEIGHT

Up to 1m (3ft 2in)

Chillies are grown in a very similar manner to tomatoes. If I had to choose for space between chillies and toms, I would favour chillies because you can just grow more of them and they store so well over winter. Chillies germinate at roughly the same temperatures as toms, but take a lot longer. They do however need to be sown very early to get a good crop. I sow mine in mid- to late winter, though you can continue to sow in early spring. They can take over two weeks to germinate, so stay patient, and they germinate much better with a propagation lid as germination is hindered if the soil surface dries out. Once up, the seedlings needs to stay a bit warmer than toms, doing best around 18–25°C (64–77°F). They are also very hungry, so will need potting

on a lot more regularly. Every time the roots poke out the bottom of the drainage holes, pot on up. You can grow chillies in pots or in the ground in a greenhouse or very sheltered spot outside. You certainly get very good heat in chillies that grow in the ground, but it does need to be very warm outside to keep most happy though there are a few reliable outdoor cropping varieties.

Prune young plants by nipping out the growing tip to promote bushy growth and more flowers. Some varieties produce something called the king chilli, which is the first fruit that appears at the first axis in the plant. If you leave this one to develop it inhibits further flowering by concentrating a lot of the growth into this single fruit, so take it off. Once flowers appear feed plants weekly

like tomatoes. The size of the plant will determine the size of pot. Dwarf ones can be grown in pots as small as 15cm (6in), but I'd err on the largest pot you can provide relative to size otherwise you will be watering like mad. Chillies need to dry out between watering. Oh, and they hate draughts at all stages of life.

Capsicum pubescens 'Rocoto Red' (also known as 'Manzano') is a mind-blowingly hot but brilliantly flavoured chilli that resembles a miniature bell pepper with distinct black seeds. Fruits start off green and mature deep red; there's also a deep purple form. It's a huge plant so needs a big pot, 15 litres (3 gallons) is good. 'Early Green Jalapeño' is one jalapeño that will ripen in British short seasons. It has the distinct corky marks on

the skin, great flavour and is perfect for pickling. 'Lemon Drop' is a lovely, tasty yellow chilli with lemon scent and flavour and the chilies dry well too. It is a big, vigorous plant, so give it a 15-litre (3-gallon) pot and you'll get loads of fruit, ripening in early autumn.

'Prairie Fire' is a dwarf variety that is super spicy, but so beautiful. The chillies start off yellow and ripen to bright red. If you don't have a greenhouse, a plant is happy on a windowsill in a 30cm (12in) pot. 'Numex Twilight' is another dwarf variety that's good for pots and highly ornamental. The fruit are upright, small and spicy. They start off purple, then turn yellow, orange and eventually red. 'Fairy Light' is a selection of 'Numex Twilight' with purple-tinged leaves.

The Toppings

Cucumbers

Cucumis sativus

POSITION *Sun* **SOIL CONDITIONS** *Organic, rich soils* **FLOWERING PERIOD** *Mid-summer onwards*
HARVESTING PERIOD *Late summer to early autumn* **HEIGHT** *50cm–3m (20in–10ft)*

There are plenty of cucumbers that can be grown outside as well as with protection. Some years I grow quite a few in pots on the patio, but my love for cucumbers is deep and I need lots of cucumbers to be happy. I long ago gave up growing very long, supermarket-style cucumbers to favour mainly Japanese, Turkish and Eastern European varieties that tend to be picked small and are perfect for pickling and for eating on the go. Their flavour is sublime.

I sow my cucumber quite late, around mid-spring. They are fast growers but utterly resent sudden cold nights or draughts so I aim to have them into the tunnel by late spring. If you are growing outside wait till late spring to sow, and plant them in the ground for early summer when the soil has truly warmed up. They germinate at 21–24°C (70–75°F) and work best if you sow the seed on its side, about 1cm (½in) deep, singly in small pots so you don't have to transplant them as they are fussy about their roots being disturbed. Grow on at 18°C (64°F) until you have four true leaves and then pot into their final larger growing pots or into beds.

I have great success using bottomless air pots on the beds in the polytunnel. These are pots with perforated sidewalls that improve air circulation and root growth. As this raises the cucumbers off the ground it leaves space around them to plant up with salads and tender herbs. Outdoors I find that the traditional method of growing cucumbers on raised mounds or ridges is best as, although they love water, they hate to sit in it. Make a mound about 10–15cm (4–6in) high,

plant the cucumber at the top and once the plant is happily growing mulch with homemade compost. I would cloche outdoor cukes until they are too tall to do so, as early fast growth is the only way to get them established.

All cucumbers need to climb up something, wires, canes or netting, but make sure it can take the weight of a fully mature plant with fruit. There's a lot said about removing side shoots for more efficient cropping, but I let mine run wild, they know what they want best. Once they start cropping, keep picking and start feeding and they'll carry on indoors until mid-autumn if you're lucky.

If you want larger fruit, I recommend 'Wautoma' which resists all known cucumber diseases and has lovely bitter-free fruit. It can grow indoors or out and be picked small for pickling or left to grow larger for slicing. 'Marketmore' is an old-fashioned cucumber with straight, dark green fruit up to 20cm (8in) long with small white spines and ridges. It's good outdoors and although the skin is quite tough by modern standards it has excellent flavour and good disease resistance. Remember that the vitamins are in the skin and learn to savour its flavour rather than peel it off.

'Crystal Lemon' is a lovely rounded lemon-shaped cuke with sweet skin which can be eaten just like an apple. 'Tamra' is another slicing cucumber that is very prolific with fantastic flavour, better off indoors that out. 'Zipangu' is a F1 Japanese hybrid (you can't collect the seed) that has spiny, dark green skin with very crisp, less watery flesh. It makes the best cucumber salad.

Basils

Ocimum basilicum

POSITION *Sun* **SOIL CONDITIONS** *Do best in organic, rich soils*
FLOWERING PERIOD *Mid-summer onward* **HARVESTING PERIOD** *Throughout the growing season*
HEIGHT *30cm (12in)* **POLYCULTURE POSITION** *Glade*

Of course you can grow basil outside – in a good summer you will grow fantastic basil outside, but in cooler summers it will be a little tough. Basil wants heat, all the time, through the night, first thing in the morning and then right the way through the day. I don't have to point out how rare that is in Britain. More than likely you will plant out great, healthy-looking young plants to find that the slugs got there first. Or that the night suddenly got a little cool and it decided to flower early or, the biggest injustice of all, that your healthy young plant has slowly shrunk in size as it shivers through the rainy cloudy days. My solution to this is to grow lots of basil, far too much, and if the summer turns out good, then there's plenty to plant out, but if doesn't I keep it in the greenhouse or a sunny kitchen windowsill and the basil is happy and so am I.

Sowing basil is simple and can be done all year round if you are growing indoors. Scatter a little seed in a seed tray or directly in the pot, don't bury the seed but let it sit on the surface of the compost and then eat the thinnings when those seedlings become overcrowded. Basil likes life in the pot, you only have to be mindful not to water late in the day as it sulks if it sits with wet feet overnight and the seedlings are then very liable to get damp off, a fungal disease which makes them wilt. As the name suggests it is caused by damp conditions, so water in the morning and allow the soil surface to dry out before watering again.

Once a basil plant flowers the leaves become less sweet, the stem becomes decidedly woody and the plant loses interest in producing more leaves but sends all its energy into creating the next generation. Don't pull up flowering basil unless you are very short on space as the flower spikes will attract pollinators if grown outside, and they are also a lovely addition to cocktails, plus the plants are self-fertile so you'll get seed which you can either collect to use for micro-green basil or soak to make basil-seed drinks which are big in Thai and Vietnamese cuisine and quite delightful.

There are numerous basil cultivars: huge-leaved forms for salads, cinnamon-, lemon- and lime-scented leaves, red, yellow and sweet green ones. You can amass quite a collection. I wouldn't live without 'Sweet Genovese', a slightly more compact, very aromatic Italian form that's perfect for pesto. 'Piccolino', sometimes sold as Greek basil, is a lovely, tiny dwarf basil that makes a neat mound of leaves and you'll need at least four plants for regular kitchen use, but it's the only one I've found that is suitable for growing outside whatever the summer in the UK. The small, highly aromatic leaves seem somewhat more slug proof than others, and it can be used to make a very neat edging plant. The lettuce-leaved Italian variety 'Foglia di Lattuga' has huge leaves the size of your hand if you grow it properly. They are mild and perfect for wrapping round a warm, vine-ripened tomato. You need to grow a single plant in a 30cm (12in) pot. If you overcrowd you just get small leaves and slug fodder, so you need to keep on top of that, but when you finally get there, it's quite something.

Additional Spaces

The Edible Water Garden

Water is an essential element in the garden, not just for thirsty roots but for the birds and other beasts that come to use your space and need a drink or a place to preen, for the beneficial insects and predators that use the water in their life cycle such as frogs, toads, dragon and hoverflies, and for tranquillity – the reflection of the sky in a still pool can be a welcome break in dense planting.

In a small garden it may seem as if there isn't space to dig a pond, but even a small pot, a bucket or container can be used to make a highly functional water garden. Many common water garden plants are edible so you can combine the best of both worlds in one space.

Most water plants need to live at a specific height within the water. Marginals don't want to be submerged and deep aquatic plants won't thrive in shallow water, nor will insects. Artificial hard-plastic ponds often come with shelves and different depths, but a homemade pond from something like a builders' bucket won't. You can adapt any space with bricks and stones to create different heights as your pond must have a margin somewhere where things can get in and out. Frogs and toads will die if they can't get out at some point and you'll drown numerous bees if you don't make sufficient exit routes. A marginal plant or even something as simple as a stick poking out from the water to the edge will help many an insect that falls in to get out again.

Arrowheads
Sagittaria species

These are some of easiest aquatics to grow as they like muddy ponds, ditches and lakes. The North American duck potato, *Sagittaria latifolia,* is delicious. Roasted roots taste somewhere between a sweet chestnut and a potato. Peel the skin after roasting as this is a little bitter tasting. *Sagittaria sagittifolia* has slightly more skinny leaves and *Sagittaria trifolia*, Chinese arrowhead, has very pointed leaves; both taste more like potatoes.

The leaves are attractive and pretty, small white flowers in summer are as attractive to ducks and waterfowl as to us, but if you give them full sun and rich aquatic compost you'll find they multiply very quickly. The tubers are best harvested in early autumn when the foliage starts to die back, leaving a few tubers to regenerate the following spring. I have found them perfectly hardy. They do best in full sun to partial shade. I've found they will grow in deep shade, but the tubers are very small.

Water Celery
Oenanthe javanic

This lovely marginal plant is a very common ingredient in large parts of Asia, particularly China and Japan. Young leaves and shoots are rich in vitamins and minerals and used much like celery, as often eaten raw or as a garnish as cooked. This plant thrives in shallow water in full sun and likes to spread so I've found its best to grow it in a large, shallow container filled with aquatic soil and topped up with water so it can expand as it needs. You could use shallow aquatic baskets, but you will have to divide plants more often. This should be done in spring.

I grow the variety *O. javanica* 'Flamingo' which has variegated pink, white and green leaves and makes a lovely addition to salads. The taste is somewhere between celery leaves and carrot leaves. There is some suggestion that too much of this plant could be toxic, but I have no idea how you'd find yourself eating enough of it!

It is not completely frost proof. I get around this by making the plant a domed cover from a clear plastic washing basket for the winter and keeping the container tucked into a corner that rarely gets frost. The plastic cloche for the container also means I get the plant growing quickly in spring.

Tsi or Fish Mint
Fish leaf, Rainbow plant
Houttuynia cordata

This is another firm favourite, not because you'll use a whole lot, but because it's a cheerful, easy-going semi-aquatic that adds an extraordinary depth to southeast Asian-inspired dishes, which you won't find in the average supermarket. *Houttuynia cordata* loves a little shade, making it perfect for corners you don't know what to do with.

It likes to grow in moist to wet soils or slightly submerged in water. It's a perfect plant for the pond margin or grown in a bucket or other watertight container. It can grow in full sun, but leaves then tend to be a bit small and tough. It grows quickly to 60cm (2ft) in flower and 1m (3ft 2in) wide, but it will tolerate being confined to smaller conditions.

There's a wonderful variety called 'Chameleon' with green, white and pink variegations. Definitely garish, it works well in a container on its own or spreading in a shady area where some cheer is needed. You can use the leaves as a garnish; they taste somewhat fishy, hence the name, but with hints of mint. It can be used in salads, often combined with coriander, vinegar, fresh chilli and soy sauce and is a classic ingredient with stir-fried beef and in many noodle dishes. Roots are edible with a fresh, spicy flavour and often cooked in southern Chinese dishes. The foliage will die back in a very hard frost, but the roots are much tougher.

Pickerel Weed
Pontederia cordata

The blue flowers of pickerel weed are a delight and it's a widely available and fully hardy deep-emergent water plant, generally grown for its ability to attract wildlife to the garden pond. It needs water 15–30cm (6–12in) deep, and it's a plant of bogs and muddy ponds so it likes quite rich aquatic soil. It spreads by thick creeping rhizomes so needs space, but if you have a container large enough (a half barrel) it's a lovely thing to grow. In a sunny spot it will flower all summer long. Its seeds are deliciously nutty when roasted, and young shoots and leaves can be used much like spinach.

The Window Box

A good windowbox for growing any sort of edible needs to be at least 20cm (8in) deep, the deeper the better, and will need to be fixed securely. Replace the soil every season and water with a liquid feed every week through the growing season to maintain vigour.

Lettuce
Lactuca sativa –
cut and come again varieties

Rocket
Eruca vesicaria subsp. *sativa*
(shade)

Thyme
Thymus species

Basil
Ocimum basilicum

Mint
Mentha species (shade) will have
to be divided every year

Perilla
Shiso, *Perilla frutescens*

Sweet violets
Viola odorata

Miner's lettuce
Claytonia perfoliata (shade)

Nasturtiums
Tropaeolum species

Chives, garlic chives
Allium moly, daffodil garlic, mouse garlic
and other *Allium* species

Strawberries
Fragaria species (semi-shade)

Chamomile
Chamemelum nobile,
Matricaria chamomilla

Eat What You Grow

The Shaded Garden

Shade is hard for many conventional annual vegetables. Tomatoes, potatoes, onions, garlic, courgettes and many herbs need full sun to do well and there is no point battling shade for these, they just won't work. Instead choose from a wider variety of woodland types that love shady spots. An edible shade garden will be a place of grazing and foraging rather than optimal production, but it can be beautiful and offer up many habitats for insects and birds.

Shrubs and Climbers

Oregon grape, *Mahonia aquifolium*, with acid yellow flowers in spring followed by tart, but delicious deep purple berries in late summer. *Mahonia japonica* 'Charity' is a highly scented option; both are good on poor urban soil
Chocolate vine, *Akebia quinata*, does well on north-facing walls and if you are very lucky you'll get soft, delicious, juicy fruit in late summer
Cultivated blackberries, *Rubus* hybrids
Caucasian spinach, *Hablitizia taminoides*
White and redcurrants, *Ribes* species

Herbaceous Layer

Landcress, *Barbarea verna*
Mistsuba or Japanese parsley, *Cyrptotaenia* japonica
Campanulas, *C. portenschlagiana, C. Poshcarkshyana*
Violets, *Viola odorata, V. labradorica*
Wild garlic, *Allium ursinum*
Victory onion, *Allium victorialis*
Wood sorrel, *Oxalis acetosella*
Solomon's Seal, *Polygonatum × hybridum*
Rhubarb, *Rheum* species
Cardamines, *Cardamine* species
Wild strawberries, *Fragaria vesca*

Index

A

achocas 53
alpine strawberry 20
angelica 86–7
anise hyssop 20
apple mint 99
apples 20, 26–9
aquilegia 99
arrowheads 182
artichokes
    Chinese artichoke 96
    globe artichokes 105
asparagus, Japanese 107

B

basil 180–1, 186
bay 71
beans 146–51
    French beans 150–1
    runner beans 148–9
beetroot 135
bellflowers 98
blackberries, cultivated 187
blueberries 20, 60–1
borage 120
brambles 74–5
broadcast sowing 22
bulbs 20
bush kales 44

C

cabbages
    annual cabbages 140–1
    tree cabbage, Jersey cabbage or
    walking stick kale 45
campanulas 20, 98, 187
caraway 89
cardamines 51, 187
cardoons 104
carrots 139
Cathay quince 36
Caucasian spinach 187

cauliflower, 'Nine Star' perennial
broccoli 46
chamomile 186
chard, Swiss 144–5
chervil 90
chillies 176–7
Chinese artichoke 96
chives 20, 186
chocolate vine 187
choy sum 165
climbers 18, 20, 72–81
    brambles 74–5
    grapes 76–7
    mashua 20, 78–9
    roses 80–1
    shaded gardens 187
coriander 90
corn salad 123
cottagers' kales 44
courgettes 156–7
crab apples 30
cuckoo flowers 51
cucumbers 170, 178–9
cucurbits 154–7
    courgettes and summer squashes
    156–7
    winter squashes 155
currants 20, 54–5, 187

D

daffodil garlic 186
damsons 41
daylilies 99
dwarf raspberry 98

E

English mace 68

F

false strawberries 99
fennel, herb 84–5
figs 38–9

fillers 116–31
    borage 120
    corn salad 123
    flowers 128–31
    landcress 126, 187
    miner's lettuce 20, 123, 186
    nasturtiums 130–1, 186
    orach, mountain spinach 118
    parsley 124
    poppies 128
    pot marigolds 129
    rocket 46, 125, 186
    salads and herbs 122–7
    tree spinach 118
    vegetable mallows 121
    watercress 127
fish leaf 184
fish mint 184
flowers 128–31
    nasturtiums 130–1, 186
    poppies 128
    pot marigolds 129
    roses 80–1
French beans 150–1
frilly mustards 142
fruit see individual types of fruit
fungus 29
fuchsia 20

G

gages 40–1
garlic chives 20, 186
glade plants 19, 20
globe artichokes 105
golden garlic 20
gooseberries 20, 56
grapes 76–7
greengages 41
greens, leafy 42–51
ground cover 92–9
    bellflowers and harebells 98
    Chinese artichoke 96

dwarf raspberry 98
marjoram 97
oregano 97
shade plants 99
strawberries 94–5, 186, 187
sun to part-shade plants 99

H
harebells 98
herb fennel 84–5
herbaceous plants 20
herbs 122–7, 168–9
   basil 180–1, 186
   bay 71
   chervil 90
   coriander 90
   English mace 68
   Korean liquorice mint 70
   marjoram 97
   mint 68–9, 186
   myrtle 71
   oregano 97
   papalo 169
   parsley 124
   perilla 168
   quillquina 169
   rosemary 66
   sage 67
   shiso 168
   summer savory 169
   sweet Nancy 68
   upright herbs 64–71
   window boxes 186
   winter savory 67
horseradish 112–13
hostas 99, 106

J
Japanese asparagus 107
Japanese ginger 110–11
Japanese medlar 36
Japanese parsley 87, 187

Japanese quince 20, 36
Jersey cabbage 45
Jerusalem artichokes 53, 109
juneberries 20, 41

K
kale 44
   annual kales 140–1
Korean liquorice mint 20, 70

L
landcress 126, 187
leafy greens 42–51
   annual kales and cabbages
      140–1
   cuckoo flowers 51
   Japanese asparagus 107
   kale, bush kales, cottager's kales
      44
   nettles 50
   'Nine Star' perennial broccoli 46
   perennial wall rocket 46
   scorzonera 49
   Solomon's seal 51, 99, 187
   sorrels 48, 99
   Swiss chard 144–5
   tree cabbage, Jersey cabbage or
      walking stick kale 45
leeks, annual 143
lemon balm 20, 99
lesser stitchwort 99
lettuce 161–3, 186
   miner's lettuce 20, 123, 186
loquat 36
lower storey plants 18, 20

M
mace, English 68
Malabar spinach 20
mallows, vegetable 121
marjoram 97
mashua 20, 78–9

medlars 34
miner's lettuce 20, 123, 186
mint 68–9, 186
   apple mint 99
   Korean liquorice mint 70
mitsuba 87, 187
mizuna 165
mountain spinach 118
mouse garlic 20, 186
mustards 166
   frilly mustards 142
myoga 110–11
myrtle 71

N
nasturtiums 130–1, 186
nettles 50
'Nine Star' perennial broccoli 46
nodding onion 20

O
oca 112
orach 118
oregano 97
Oregon grape 187
ornamental edibles 100–15
   cardoons 104
   globe artichokes 105
   horseradish 112–13
   hostas 106
   Japanese asparagus 107
   Jerusalem artichokes 109
   myoga, Japanese ginger 110–11
   oca 112
   rhubarb 114–15
   sedums 102–3
   tulips 115

P
pak choi 164
   rosette pak choi 164
papalo 169

Index

parsley 124

parsnips 138

pears 31–3

peas 146, 152–3

pepper trees 62–3

perennial wall rocket 46

perilla 168, 186

pickerel weed 184

plums 40–1

pollinator plants 100–15

  cardoons 104

  globe artichokes 105

  horseradish 112–13

  hostas 106

  Japanese asparagus 107

  Jerusalem artichokes 109

  myoga, Japanese ginger 110–11

  oca 112

  rhubarb 114–15

  sedums 102–3

  tulips 115

polyculture

  definition of 8–11

  why grow your own food 12–15

poppies 128

pot marigolds 129

potted gardens 20

Q

quillquina 169

quinces 35

R

radishes 137

rainbow plant 184

raspberries 58–9

  dwarf raspberry 98

redcurrants 20, 187

rhubarb 114–15, 187

rocket 125, 186

  perennial wall rocket 46

root vegetables 134–9

  beetroot 135

  carrots 139

  parsnips 138

radishes 137

turnips 136

rosemary 66

roses 80–1

rosette pak choi 164

runner beans 148–9

S

sage 67

salads 122–7, 160–7

  choy sum 165

  corn salad 123

  landcress 126, 187

  lettuces 161–3

  miner's lettuce 20, 123, 186

  mizuna 165

  mustards 166

  pak choi 164

  rocket 125

  rosette pak choi 164

  watercress 127

savory

  summer savory 169

  winter savory 67

scorzonera 49

sedums 102–3

self-fertile trees 28

serviceberry 41

shaded gardens 187

  deep shade plants 18

  herbaceous layer 187

  light shade plants 18

shiso 168, 186

shrubs 20, 52–63

  blueberries 60–1

  currants 20, 54–5, 187

  gooseberries 56

  pepper trees 62–3

  raspberries 58–9

  shaded gardens 187

Solomon's seal 51, 99, 187

sorrel 48, 99

sowing, broadcast 22

squashes

  climbing 53

summer squashes 156–7

winter squashes 155

strawberries 94–5, 186

  wild strawberries 187

summer savory 169

summer squashes 156–7

sun, full sun plants 19

sweet cicely 89

sweet Nancy 68

sweet violets 20, 186

Swiss chard 144–5

T

tender things 170–81

  basil 180–1, 186

  chillies 176–7

  cucumbers 170, 178–9

  tomatoes 170, 171–5

thyme 186

tomatoes 53, 170, 171–5

toppings 132–81

  annual kales and cabbages 140–1

  annual leeks 143

  basil 180–1, 186

  beetroot 135

  carrots 139

  chillies 176–7

  choy sum 165

  courgettes and summer squashes 156–7

  cucumbers 170, 178–9

  cucurbits 154–7

  French beans 150–1

  frilly mustards 142

  herbs 122–7, 168–9

  lettuces 161–3, 186

  mizuna 165

  mustards 166

  pak choi 164

  papalo 169

  parsnips 138

  peas and beans 146–53

  perilla 168

  quillquina 169

radishes 137
root vegetables 134–9
rosette pak choi 164
runner beans 148–9
salads 122–7, 160–7
shiso 168
summer savory 169
Swiss chard 144–5
tender things 170–81
tomatoes 170, 171–5
turnips 136
winter squashes 155
tree cabbage 45
tree spinach 118
trees 20, 24–41
apples 26–9
Cathay quince 36
crab apples 30
figs 38–9
Juneberry, serviceberry 41
loquat or Japanese medlar 36
medlars 34
pears 31–3
plums and gages 40–1
quinces 35
tsi 184
tulips 115
turnips 136

U
umbels 82–91
angelica 86–7
caraway 89
chervil 90
coriander 90
herb fennel 84–5
mitsuba, Japanese parsley 87
sweet Cicely 89
wild Korean celery 88
understorey plants 19
upper storey plants 18, 20

V
vegetable mallows 121
vegetables

root vegetables 134–9
see also individual types of
vegetable
victory onion 187
violets 99, 187

W
walking stick kale 45
wall rocket, perennial 46
water celery 184
water gardens, edible 182–5
watercress 127
Welsh onion 20
white currants 20
wild garlic 99, 187
wild Korean celery 88
wild strawberries 187
window boxes 186
winter savory 67
winter squashes 155
wood sorrel 99, 187

ACKNOWLEDGEMENTS

Thank you to Anka Dabrowksa for all the fine points to her beautiful drawings, to Roo for all the magic in every shot, to Sophie Allen for bringing this book into the world and Judith for steering us home, to Charlie for careful edits and to Carol for her flawlessly good style. Finally, to Ele, the small dog and the garden, you three make my world.

UK/US GLOSSARY

allotment – community garden
aubergine – eggplant
beetroot – beets
broad beans – fava beans
coriander – cilantro
courgette – zucchini
dustbin – garbage can
mangetout – snow pea
pavement – sidewalk
rocket – arugula